Advanced Applied Mathematics

Ram Bilas Misra

CWP
Central West Publishing

Advanced Applied Mathematics

by

Prof. Dr. Ram Bilas Misra

Ex Vice Chancellor, Avadh University, Faizabad, U.P. (India);

Professor of Mathematics, Lebanese–French University, Erbil, Kurdistan (Iraq) – designate.

Former: *Dean*, Faculty of Science, A.P. Singh University, Rewa, M.P. (**India**);
Prof., Dept. of Maths., Higher College of Edn., Aden Univ., Aden (**Yemen**);
Professor & Head, Dept. of Maths. & Stats., A.P.S. University, Rewa, M.P. (**India**);
Prof., Dept. of Maths., College of Science, Salahaddin University, Erbil (**Iraq**);
UGC Visiting Prof., Mahatma Gandhi Kashi *Vidyapith*, Varanasi, U.P. (**India**);
Professor, Dept. of Maths, Ahmadu Bello Univ., Zaria (**Nigeria**) – designate;
Prof. & Head, Dept. of Maths. & Comp. Sci., Univ. of Asmara, Asmara (**Eritrea**);
Director, Unique Inst. of Business & Technol., Modi Nagar, Ghaziabad, U.P. (**India**);
Prof. & Head, Dept. of Maths., Phys. & Stats., Univ. of Guyana, Georgetown (**Guyana**);
Prof. & Head, Dept. of Maths., Eritrea Inst. of Technology, Mai Nefhi (**Eritrea**);
Prof.& Head, Dept. of Maths., School of Engg., Amity Univ., Lucknow, U.P. (**India**);
Prof. & Head, Dept. of Maths. & Comp. Sci., PNG Univ. of Technology, Lae (**PNG**);
Prof. of Maths., Teerthankar Mahaveer University, Moradabad, U.P. (**India**);
Prof., Dept. of Maths, Oduduwa Univ., Ipetumodu, Osun State (**Nigeria**) – designate;
Prof., Dept. of Maths, Adama Science & Technology Univ., Adama (**Ethiopia**);
Prof. & Head, Dept. of Maths. & C.S., Bougainville Inst. of Bus. & Tech., Buka (**PNG**) – designate;
Prof. & Head, Dept. of Maths., J.J.T. University, Jhunjhunu, Rajasthan (**India**);
Dean, Faculty of Science, J.J.T. University, Jhunjhunu, Rajasthan (**India**);
Professor, Dept. of Maths, Wollo University, Dessie, Wollo (**Ethiopia**);
Professor, Dept. of Appld. Maths., State Univ. of New York, Incheon (**S. Korea**)
Prof., Dept. of Maths. & Computing Sci., Divine Word Univ., Madang (**PNG**);
Director, Maths., School of Sci. & Engg., Univ. of Kurdistan Hewler, Erbil (**Iraq**);
DAAD Fellow, University of Bonn, Bonn (**Germany**);
Visiting Professor, University of Turin, Turin (**Italy**);
Visiting Professor, University of Trieste, Trieste (**Italy**);
Visiting Professor, University of Padua, Padua (**Italy**);
Visiting Professor, International Centre for Theoretical Physics, Trieste (**Italy**);
Visiting Professor, University of Wroclaw, Wroclaw (**Poland**);
Visiting Professor, University of Sopron, Sopron (**Hungary**);
Reader, Dept. of Maths. & Stats., South Gujarat University, Surat, Gujarat (**India**);
Reader, Dept. of Maths. & Stats., University of Allahabad, Allahabad, U.P. (**India**);
Asst. Prof., Dept. of Maths., College of Sci., Mosul Univ., Mosul (**Iraq**) – designate;
Senior most *NCC Officer* (Naval Wing), Univ. of Allahabad, Allahabad, U.P. (**India**);
Lecturer, Dept. of Maths., KKV Degree College, Lucknow, U.P. (**India**)

2018

This edition has been published by Central West Publishing, Australia

For more information about the books published by Central West Publishing, please visit https://centralwestpublishing.com

NATIONAL LIBRARY OF AUSTRALIA

A catalogue record for this book is available from the National Library of Australia

ISBN (print): 978-1-925823-11-0

DEDICATED TO

MY TEACHERS ESPECIALLY

Prof. Dr. R.S. Mishra

Prof. K.S. Shukla

Padmashri Prof. Dr. Ratna Shankar Mishra,

M.Sc., Ph.D., D.Sc., FNA, FNASc, FASc, FIAPSc, President (Maths. Section),
General-Secretary & General President of Indian Science Association,
and Founder Patron,
International Academy of Physical Sciences, Allahabad (India)
(10.08.1918* – 23.08.1999), *Actual 26.9.1919/*Vijayadashmi*

Professor & Head, Gorakhpur, Allahabad & Banaras Hindu Universities (India);
Visiting Prof. to Indiana Univ., Bloomington (USA);
Univ. of Waterloo (Canada); Univ. of Windsor (Canada);
Kuwait Univ. (Kuwait); Turin Univ. (Italy); Univ. of Jammu (India), etc.

Ex Vice-Chancellor, CSM Kanpur University, Kanpur (India);
and Lucknow University, Lucknow (India)

AND

Professor Dr. Kripa Shankar Shukla

M.A. (Alld., India), D.Litt. (Lucknow, India)
(10.7.1918–22.9.2007)

IN THEIR CENTENARY YEAR

What a coincidence of both being self-made & having lost their fathers early.

In Memoriam

Prof. Franco Fava (26.1.1922 - 19.4.2018)

Professor Dr. Franco Fava

Former Director, Dept. of Mathematics, Univ. of Turin, Turin (Italy);
Emeritus Professor, Institute of Geometry & Topology, Univ. of Turin (Italy);
Professor of Mathematics, Politecnico di Torino, Turin (Italy);
Member, Accademia delle Scienze di Torino, Turin (Italy).

CONTENTS

PREFACE

Idea of writing this book has roots in an invitation to the author extended by Dayanand Science College (DSC), Latur, Maharashtra (India) to deliver lectures in a Refresher Course for the College Teachers during June, 2018. The material in the book covers the topics needed for an advanced course in Applied Mathematics taught in higher institutions. Some of the topics especially on Multi-valued Functions, Fourier Transforms, Diffusion Equations and Non-Relativistic Scattering are relatively newer. It offers one year/two semester course suitable for graduate students with 3 hours weekly credits.

The subject matter is presented here in fifteen chapters including the first two dealing with prerequisites and ordinary differential equations. Chapters are divided into Sections and the discussion within the Sections is presented in the form of Definitions, Theorems, Corollaries, Notes and Examples. Most of the chapters end in a Problem Set containing unsolved exercises, but solution to challenging ones is provided with necessary hints. The sub-titles within the Sections are numbered in decimal pattern. For instance, the equation number (c.s.e) refers to the eth equation in the sth section of Chapter c. When c coincides with the chapter at hand, it is dropped. Adequate references to the results appeared earlier are made in the text avoiding their unnecessary repetition. Double slashes marked at the end of Theorems, Corollaries, Solutions of Exercises, etc., indicate their completion.

My long teaching career of more than ***five decades*** at various universities round the globe and research expertise in different fields helped me for lucid presentation of the subject. In the preparation of the text, I have immensely benefited through books (listed in the References), personal interactions with my former colleagues: Prof. S.N. Pandey, Dr. S.N. Mishra of Dr. A.P.J. Abdul Kalam Technical University, Lucknow (India) and ex-student Dr. Satgur Prasad Khare of Delhi (India). Perhaps it is the divine will that the organizers of the course provided this opportunity to me in this advanced age to offer the homage to my Ph.D. supervisor (late) Professor Dr. R. S. Mishra (at University of Allahabad, India) in his centenary year. Hence, the book is dedicated to him and my teachers only. Incidentally, it is also the centenary year of (one of the best) teachers, Prof. Dr. Kripa Shankar Shukla, who taught so well at Lucknow University, India. His deep commitment to his job suddenly flashed in my mind and reminded me not to be missed.

My thanks are also due to various Universities all over the world especially University of Allahabad (India); University of Guyana (Georgetown); P.N.G. University of Technology (Lae, Papua New Guinea); Adama Science & Technology University (Adama, Ethiopia); State University of New York (Incheon, South Korea); Divine Word University (Madang, P.N.G.), etc., where I gained a lot while exposing my expertise. Organizers of the Refresher Course at DSC, Latur also deserve my gratitude, who honoured me to have selected as a resource person clubbing me with some of the best contemporary mathematicians of the country, especially Professor V. Kannan. It may go unfair on my part if I do not record the sincere cooperation of my family especially the better-half (Mrs. Rekha Misra), whom I often had to exhaust for my academic passion. During preparation of the book, I lost my most humble elder colleague (Prof. Dr. Franco Fava of Turin, Italy), who had been my host there for over a dozen times. I pray for his eternal peace and enough strength to the bereaved family and friends to mourn his irreparable loss forever. Sincere thanks are also due to the publisher for their valuable cooperation and bringing the book into limelight in a limited time.

Although proofs are read with utmost care and solutions to problems are verified repeatedly, yet an oversight or any discrepancy brought to the notice of the author by the inquisitive readers(s) shall be thankfully acknowledged. What a surprising coincidence of completing the proof-corrections on the day when I lost my mother in 2003.

Lucknow (India): August 21, 2018 Ram Bilas Misra

CHAPTER 1

PRE-REQUISITES

I. Real Number System

§ 1. Natural Numbers

Counting numbers (more precisely natural numbers, i.e. positive integers) is prevalent from immemorial times. In the early stages, Nomadic people used to count in terms of stone pieces or their cattle.

Systematic study of natural numbers began with the postulates of G. Peano, who gave the following five postulates:

(i) 1 is a natural number;
(ii) given a natural number a, its successor a^+ ($= a + 1$) is also a natural number;
(iii) 1 is not the successor of any natural number;
(iv) given two natural numbers a and b, if their successors are equal so are the natural numbers;
(v) if P is a set containing 1 and for any natural number a in P, there also exists its successor in P, then P is the complete set of natural numbers.

The set of natural numbers is denoted by N:

$$\mathrm{N} = \{1, 2, 3, \ldots\}. \tag{1.1}$$

The set N has no upper bound, i.e. there is no last natural number.

§ 2. Integers

The natural numbers, their negatives and zero form the whole numbers called *integers*. The set of integers is denoted by Z:

$$\mathrm{Z} = \{\ldots, -3, -2, -1, 0, 1, 2, 3, \ldots\} \tag{2.1}$$

This set has neither least nor upper bound.

2.1. Odd and even numbers: ± 1, ± 3, ± 5, … are called odd numbers, while 0, ± 2, ± 4, … as even numbers.

§ 3. Rational Numbers

The fractional numbers of the form p/q, where p is any integer, but q is a non-zero integer are called *rational numbers*.

On the other hand, the numbers, which cannot be expressed as above, are called *irrationals*. For instance $\sqrt{2}$, $\sqrt{3}$, $\sqrt{5}$, e, π etc., are irrationals.

Note 3.1. The rational numbers and irrationals together constitute the real numbers. The set of real numbers is denoted by $\mathbb{R}$.

3.1. Approximate values of some irrational numbers

$$\sqrt{2} \approx 1.4142, \qquad \sqrt{3} \approx 1.732, \qquad \pi \approx 3.1416.$$

§ 4. Prime and Composite Numbers

An integer m with only factors ± 1 and ± m is called a *prime* number. Thus, the first prime number is 2. The largest prime number accounted so far by a German medical doctor Dr. Martin Nowak on 18th February, 2005 is $2^{25964951} - 1$ having 7816230 digits.

4.1. An integer other than a prime number is called a *composite number*. For instance, 4 is a composite number.

4.2. Relatively prime numbers: Two integers having no common (integral) factor(s) other than ± 1 are called *relatively prime*. For example, 4 and 9 are relatively prime numbers. The numbers 2 and 3 are individually primes as well as relatively prime numbers.

§ 5. Algebraic and Transcendental Numbers

A real number seen as a root of some algebraic equation (involving a polynomial in some variable say x) is called an *algebraic number*. On the other hand, if a number cannot be expressed as a root of some algebraic equation, it is called a *transcendental number*.

Example 5.1. Trigonometric functions have transcendental values.

II. Complex Numbers

§ 1. Complex Number

A number of the form $x + iy$, where x, y are real numbers, whereas $i = \sqrt{(-1)}$ is an imaginary number, is called a *complex number*. It is usually denoted by z:

$$z = x + iy \tag{1.1}$$

with its real part $\text{Rl}(z) = x$ and imaginary part $\text{Im}(z) = y$.

The numbers z and

$$\bar{z} = x - iy \tag{1.2}$$

are complex conjugates of each other. Addition and subtraction of these conjugates also yield

$$(z + \bar{z})/2 = \text{Rl}(z) = x, \tag{1.3}$$

and

$$(z - \bar{z})/2i = \text{Im}(z) = y. \tag{1.4}$$

§ 2. Algebraic Laws of Complex Numbers

Given two complex numbers $z_1 = x_1 + iy_1$ and $z_2 = x_2 + iy_2$, their sum, difference, multiplication and division are given by

$$z_1 + z_2 = (x_1 + iy_1) + (x_2 + iy_2) = (x_1 + x_2) + i(y_1 + y_2), \tag{2.1}$$

$$z_1 - z_2 = (x_1 + iy_1) - (x_2 + iy_2) = (x_1 - x_2) + i(y_1 - y_2), \tag{2.2}$$

$$z_1 . z_2 = (x_1 + iy_1).(x_2 + iy_2) = (x_1 x_2 - y_1 y_2) + i(x_1 y_2 + x_2 y_1), \tag{2.3}$$

$$z_1 / z_2 = (x_1 + iy_1)/(x_2 + iy_2) = (x_1 + iy_1)(x_2 - iy_2)/(x_2 + iy_2)(x_2 - iy_2)$$

$$= \{(x_1 x_2 + y_1 y_2) + i(x_2 y_1 - x_1 y_2)\}/(x_2^2 + y_2^2). \tag{2.4}$$

The numbers so obtained are also complex numbers in general.

Note 2.1. The division of a complex number by another complex number is defined only when the divisor z_2 is non-zero, i.e.

$$x_2^2 + y_2^2 > 0.$$

§ 3. Modulus and Argument of z

Employing the polar coordinates (r, θ) defined by

$$x = r\cos\theta \qquad \text{and} \qquad y = r\sin\theta, \tag{3.1}$$

so that

$$r = \sqrt{(x^2 + y^2)} \qquad \text{and} \qquad \theta = \tan^{-1}(y/x). \tag{3.2}$$

A complex number can be alternately expressed as

$$z = x + iy = r(\cos\theta + i\sin\theta) = re^{i\theta}. \tag{3.3}$$

Thus, the real number r and angle θ as given by (3.2) define the modulus and argument of z respectively. We write

$$\text{mod } z = |z| = r = \sqrt{(x^2 + y^2)}, \tag{3.4}$$

and

$$\arg z = \theta = \tan^{-1}(y/x). \tag{3.5}$$

The argument of z is also called the *amplitude* of z. It has infinitely large number of values of which the one lying between $-\pi$ and π is called the *principal value* of θ.

§ 4. Properties of Complex Numbers

4.1. The relations (1.3) and (1.4) also yield

$$\text{Rl}(z) + i\,\text{Im}(z) = z. \tag{4.1}$$

4.2. $$|z| = \sqrt{\{(\text{Rl}(z)^2 + \text{Im}(z)^2\}} = |\bar{z}|. \tag{4.2}$$

4.3. $$z\bar{z} = (x + iy)(x - iy) = x^2 + y^2 = |z|^2 = |\bar{z}|^2. \tag{4.3}$$

4.4. $$\overline{z_1 + z_2} = (x_1 + x_2) - i(y_1 + y_2) = (x_1 - iy_1) + (x_2 - iy_2) = \bar{z}_1 + \bar{z}_2. \tag{4.4}$$

4.5. $$\overline{z_1 z_2} = (x_1 x_2 - y_1 y_2) - i(x_1 y_2 + x_2 y_1)$$

$$= x_1(x_2 - iy_2) - iy_1(x_2 - iy_2) = (x_1 - iy_1)(x_2 - iy_2) = \bar{z}_1\bar{z}_2. \tag{4.5}$$

4.6. $$\overline{z_1 / z_2} = \{(x_1 x_2 + y_1 y_2) - i(x_2 y_1 - x_1 y_2)\}/(x_2^2 + y_2^2)$$

$$= \{(x_1 - iy_1)\, x_2 + iy_2\, (x_1 - iy_1)\} \,/\, (x_2^2 + y_2^2)$$

$$= (x_1 - iy_1)\,(x_2 + iy_2) \,/\, (x_2 + iy_2)\,(x_2 - iy_2) = (x_1 - iy_1) \,/\, (x_2 - iy_2) = \bar{z}_1 / \bar{z}_2 \qquad (4.6)$$

provided $\bar{z}_2 \neq 0$.

Example 4.1. Reduce the complex number $1 - \cos\alpha + i \sin\alpha$ to the modulus and amplitude form.

Solution. Putting $1 - \cos\alpha = r\cos\theta$ and $\sin\alpha = r\sin\theta$, we find

$$r^2 = (1 - \cos\alpha)^2 + \sin^2\alpha = 2 - 2\cos\alpha = \{2\sin(\alpha/2)\}^2$$

$\Rightarrow$

$$r = 2\sin(\alpha/2),$$

and

$$\tan\theta = \sin\alpha \,/\, (1 - \cos\alpha) = \cot(\alpha/2) = \tan(\pi/2 - \alpha/2)$$

$\Rightarrow$

$$\theta = (\pi - \alpha)/2$$

Hence, the given number reduces to

$$r(\cos\theta + i\sin\theta) = 2\sin(\alpha/2)\,[\cos\{(\pi - \alpha)/2\} + i\sin\{(\pi - \alpha)/2\}]. \;//$$

§ 5. Geometrical Representation of Complex Numbers

Let X and Y denote the sets of all real (respectively imaginary) numbers along the real axis $x'Ox$ (i.e. the x-axis extending from $-\infty$ to ∞) and the imaginary axis $y'Oy$ (i.e. the y-axis also extending from $-\infty$ to ∞). Let any point P has rectangular Cartesian coordinates (x, y) or the complex coordinate $z = x + iy$. The plane containing above axes is called the *Argand plane* (or *Argand diagram*). Since a point on y-axis is imaginary (i.e. i^{th} multiple of a real number), it is denoted by iy.

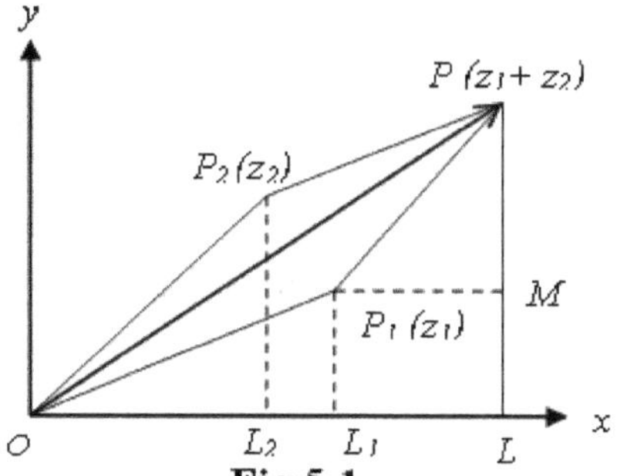

Fig 5.1

Also, comparing the mod (z) with the magnitude of a vector $\overrightarrow{OP}$ we note that the complex number z behaves as the vector $\overrightarrow{OP}$:

$$\overrightarrow{OP} = z. \tag{5.1}$$

5.1. Geometrical Representation of $z_1 + z_2$:

Let P_1 ($z_1 = x_1 + iy_1$) and P_2 ($z_2 = x_2 + iy_2$) be two points on the Argand plane. We complete the parallelogram OP_1PP_2 and draw perpendiculars P_1L_1, P_2L_2, PL to x-axis and P_1M to PL. The Cartesian coordinates of P are (OL, LP):

$$OL = OL_1 + L_1L = x_1 + P_1M = x_1 + OL_2 = x_1 + x_2,$$

$$LP = LM + MP = L_1P_1 + L_2P_2 = y_1 + y_2.$$

Therefore, the complex coordinate z of P is

$$z = OL + i\,(LP) = (x_1 + x_2) + i\,(y_1 + y_2) = (x_1 + iy_1) + (x_2 + iy_2) = z_1 + z_2. \tag{5.2}$$

Thus, $z_1 + z_2$ represents the vector along the diagonal $\overrightarrow{OP}$ of above parallelogram formed by z_1 and z_2.

5.2. Geometrical Representation of $z_1 - z_2$:

Extending $\overrightarrow{P_2O}$ up to P_2' so that $\overrightarrow{OP_2'} = -z_2$. We complete the parallelogram OQP_1P_2. Therefore,

$$\overrightarrow{P_1Q} = -z_2 \Rightarrow \overrightarrow{OQ} = \overrightarrow{OP_1} + \overrightarrow{P_1Q} = z_1 - z_2,$$

which is also equal to $\overrightarrow{P_2P_1}$. Thus, $z_1 - z_2$ represents a vector along the side P_2P_1 of the triangle OP_2P_1.

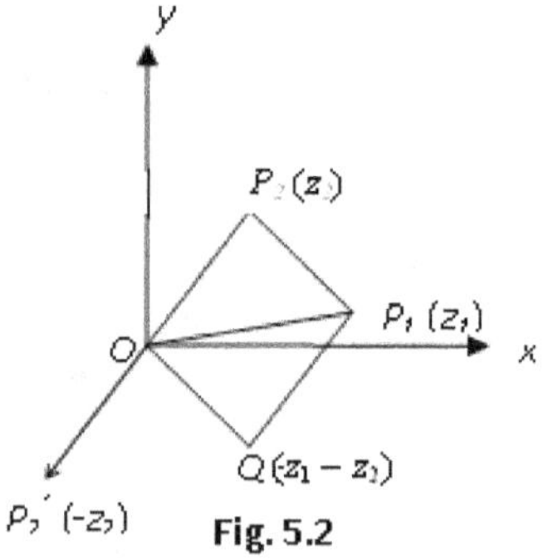

Fig. 5.2

Theorem 5.1. Given any two complex numbers z_1, z_2 there hold the relations:
(*triangular property*) $$|z_1 + z_2| \le |z_1| + |z_2|, \tag{5.3}$$

and

$$|z_1 - z_2| \ge |z_1| - |z_2|. \tag{5.4}$$

Proof. (i) As per Fig. 5.1, there holds the triangular property:

$$OP < OP_1 + P_1 P, \qquad \text{i.e.} \qquad |z_1 + z_2| < |z_1| + |z_2|.$$

If, however, the points O, P_1, P_2 are collinear there holds the equality:

$$OP = OP_1 + P_1 P, \qquad \text{i.e.} \qquad |z_1 + z_2| = |z_1| + |z_2|.$$

Thus, (5.3) is established in either case.

(ii) Again, writing z_1 as $(z_1 - z_2) + z_2$, we have

$$|z_1| = |(z_1 - z_2) + z_2| \leq |z_1 - z_2| + |z_2|, \qquad \text{by (5.3).}$$

This implies (5.4).

Note 5.1. In general, there holds

$$|z_1 + z_2 + \ldots + z_n| \leq |z_1| + |z_2| + \ldots + |z_n|. \tag{5.5}$$

Theorem 5.2. There hold the relations

$$|z_1 . z_2| \leq |z_1| . |z_2|, \tag{5.6}$$

$$\text{amp}\,(z_1 . z_2) \leq \text{amp}\,(z_1) + \text{amp}\,(z_2). \tag{5.7}$$

Proof. Writing two complex numbers z_1, z_2 in their polar forms (3.3):

$$z_1 = r_1 (\cos \theta_1 + i \sin \theta_1) = r_1 e^{i\theta_1}, \quad z_2 = r_2 (\cos \theta_2 + i \sin \theta_2) = r_2 e^{i\theta_2},$$

we derive

$$z_1 . z_2 = (r_1 r_2)\, e^{i(\theta_1 + \theta_2)}, \tag{5.8}$$

so that

$$|z_1 . z_2| = r_1 r_2 = |z_1| . |z_2|,$$

and

$$\text{amp}\,(z_1 . z_2) = \theta_1 + \theta_2 = \text{amp}\,(z_1) + \text{amp}\,(z_2),$$

as

$$|e^{i(\theta_1 + \theta_2)}| = |\cos(\theta_1 + \theta_2) + i \sin(\theta_1 + \theta_2)| = 1.$$

Example 5.1. The equation $| z | = 1$ represents a *unit circle* centered at the origin.

Solution. Writing z as (1.1) and using (3.4) for its modulus the given equation yields $x^2 + y^2 = 1$ representing a unit circle with centre at origin. //

§ 6. Logarithm of a Complex Number

Theorem 6.1. The natural logarithm of a complex number z given by (1.1) is

$$\log_e (x + i\,y) = \log_e \sqrt{(x^2 + y^2)} + i \tan^{-1} (y / x), \qquad (6.1a)$$

or, equivalently

$$\ln z = \ln | z | + i \arg (z). \qquad (6.1b)$$

Proof. The polar form of the number is given by (3.3). Taking its natural logarithm we, thus, obtain

$$\log_e z = \log_e r + i\,\theta.$$

The same, for (3.4), (3.5), assumes the forms (6.1). //

Note 6.1. For the principal value of arg z (which lies between $-\pi$ to π), the relations (6.1) determine the principal value of ln z. Its general value is given by

$$\text{Ln } z = \ln | z | + 2n\pi\, i. \qquad (6.2)$$

Corollary 6.1. $$\ln (\pm i) = \pm \pi i / 2. \qquad (6.3)$$

Proof. Writing

$$\pm i = \cos (\pi/2) \pm i \sin (\pi/2) = e^{\pm \pi i / 2},$$

and taking its natural logarithm we immediately get (6.3). //

Corollary 6.2. $$\ln (-1) = \pm \pi\, i. \qquad (6.4)$$

Proof. Since $-1 = \cos \pi + i \sin \pi = e^{\pm \pi i}$, the natural logarithm of -1 is as above. //

Note 6.2. Logarithm of a negative real number is always imaginary.

Example 6.1. Show that

$$(i)^{i} = e^{-\pi/2}. \tag{6.5}$$

Solution. Let $(i)^i = a + i\,b$. Taking its natural logarithm and putting from (6.1a) we get

$$i \ln i = \ln(a + i\,b) = \ln \sqrt{(a^2 + b^2)} + i \tan^{-1}(b/a).$$

Putting from (6.3) and equating real and imaginary parts we, thus, have

$$\tan^{-1}(b/a) = 0 \quad \Rightarrow \quad b = 0 \text{ and } -\pi/2 = \ln a \quad \Rightarrow \quad a = e^{-\pi/2}.//$$

III. Algebraic Group

§ 1. Some Algebraic Structures

A non-empty set G with an operation ○ satisfying some of the algebraic laws forms an algebraic structure. Some of the simple algebraic structures are the following ones:

1.1. Monoid: The structure (G, ○) is called a *monoid*, if ○ is only binary in G.

1.2. Groupoid (or semigroup): If ○ is both binary and associative in G, then (G, ○) is called a *groupoid* or *semigroup*.

1.3. Semigroup with identity: If there also holds the identity law with respect to a binary and associative operation ○ in a non-empty set G then (G, ○) is called a *semi-group with identity*.

1.4. Commutative semigroup: If there hold binary, associative and commutative laws w.r.t. the operation ○ in G, then (G, ○) is called a *commutative semigroup*.

1.5. Group: If ○ satisfies binary, associative, identity and inverse laws in a non–empty set G, then (G, ○) is called an *algebraic group*.

1.6. Commutative (or abelian) group: If o satisfies the first five algebraic laws given above in a non-empty set G, then (G, ○) is called a *commutative* (or *abelian*) *group*.

Example 1.1. (E, +), (Z, +), (Q, +), (R, +) and (C, +) are all algebraic groups.

Example 1.2. The singleton sets {0} and {1} form groups w.r.t. addition and multiplication respectively.

Example 1.3. The sets of non-zero rational numbers, non-zero real numbers and non-zero complex numbers also form groups w.r.t. multiplication.

Example 1.4. The set of cube roots of unity forms a group w.r.t. multiplication.

Solution. Denoting the cube roots of unity (i.e. 1) by 1, ω, ω^2 their set

$$G = \{1, \omega, \omega^2\} \tag{1.1}$$

satisfies all the four postulates in the definition of a group. The elements ω and ω^2 are inverses of each other. //

Note 1.1. All above groups are indeed *abelian* groups.

Example 1.5. The set of 2×2 non-zero matrices defined over the set of real (or complex) numbers forms a group w.r.t. matrix multiplication. However, it is not commutative in general.

§ 2. Some Properties of Groups

Definition 2.1. An element of an algebraic structure G is called *unit* if its inverse (w.r.t. binary operation $\circ$ in G) also exists in G.

Example 2.1. The element 1 is unit in the semigroup $(N, \cdot)$.

Theorem 2.1. The units of a semi–group $(G, \circ)$ with *identity* form a group w.r.t. the same binary operation $\circ$.

Proof. Let $(G, \circ)$ be a semigroup with identity and H be the set of the units of the semi-group. All the elements of H are also the elements of G, where $\circ$ is binary, associative and satisfies the identity law as per the hypothesis. Further, as per above definition every element of H possesses its inverse (w.r.t. $\circ$) in H itself. Thus, all the four postulates of a group are satisfied by $\circ$ in H making $(H, \circ)$ as a group. //

Theorem 2.2. A semigroup (G, $\circ$) becomes a group if the equations

$$a \circ x = b \quad \text{and} \quad y \circ a = b \tag{2.1}$$

are uniquely soluble in G $\forall$ $a, b \in$ G.

Proof. As per definition of a semigroup the operation o is binary and associative in a non-empty set G. Let the equations (2.1) be soluble in G. Replacing b by a only the equation

$$a \circ x = a \qquad \text{(respectively)} \qquad y \circ a = a$$

implies the existence of right (resp. left) identity element w.r.t. $\circ$ in G. Both identities are equal, say e, for the uniqueness of the solution.

Next, replacing b by e, the equation

$$a \circ x = e \qquad \text{(respectively)} \qquad y \circ a = e$$

implies the existence of right (resp. left) inverse of a w.r.t. $\circ$ in G. Further, let the inverse of a be denoted by a^{-1} being unique, which is unique. Thus, all the four postulates of a group are satisfied making (G, $\circ$) a group.

Conversely, if (G, $\circ$) is a group the equations (2.1) possess solutions

$$x = a^{-1} \circ b \quad \text{and} \quad y = b \circ a^{-1}$$

as they satisfy the respective equations:

$$a\circ(a^{-1}\circ b) = (a\circ a^{-1})\circ b, \quad \text{by associative law holding in group (G, } \circ)$$

$$= e \circ b \;\; \text{(by inverse law)} = b \;\; \text{(by identity law);}$$

and similarly,

$$(b\circ a^{-1})\circ a = b\circ(a^{-1}\circ a) = b\circ e = b. //$$

Note 2.1. Both the identity element and the inverse element of a given element in a group are unique.

Definition 2.2. A subset H of a group (G, $\circ$) forms a *subgroup* of (G, $\circ$) if (H, $\circ$) itself is a group.

Example 2.2. The following groups form a chain of subgroups of respective groups:

$$(\{0\}, +) \subseteq (E, +) \subseteq (Z, +) \subseteq (Q, +) \subseteq (R, +) \subseteq (C, +).$$

Example 2.3. The following groups are related to each other under the relation of being a subgroup:

$$(\{1\}, \cdot) \subseteq (Q^*, \cdot) \subseteq (R^*, \cdot) \subseteq (C^*, \cdot).$$

Example 2.4. The square roots of unity form a subgroup of a group of the fourth roots of unity w.r.t. multiplication.

Solution. The square roots of unity are 1, – 1 and the fourth roots of unity are 1, i, – i, – 1, where $i = \sqrt{(-1)}$ is an imaginary number. Clearly, the set of square roots is a proper subset of the set of the fourth roots. Furthermore, both the sets {1, – 1} and {1, i, – i, – 1} form groups w.r.t. multiplication. Hence, we have the statement. //

Theorem 2.3. A subset H of a group (G, $\circ$) forms a subgroup of (G, $\circ$) if $\forall\ a, b\ \varepsilon\ H \ \exists \ \ a \circ b^{-1}\ \varepsilon\ H$.

Proof. Taking (H, $\circ$) as a subgroup of (G, $\circ$) the necessary part of the theorem is easily established. (H, $\circ$) itself being a group

$$a, b\ \varepsilon\ H \ \Rightarrow\ a, b^{-1}\ \varepsilon\ H \Rightarrow a \circ b^{-1}\ \varepsilon\ H.$$

Conversely, beginning with the condition and taking $b = a$, we conclude the existence of the identity element in H:

$$a, a\ \ \varepsilon\ H \ \Rightarrow\ a \circ a^{-1} = e\ \varepsilon\ H.$$

Next, applying the condition for e, $a\ \varepsilon$ H so that $e \circ a^{-1} = a^{-1}\ \varepsilon$ H, the existence of the inverse of any element a in H is concluded.

Finally, applying the condition for elements a, b^{-1} of H:

$$a, b\ \varepsilon\ H \ \Rightarrow\ a, b^{-1}\ \varepsilon\ H \qquad \Rightarrow \qquad a \circ (b^{-1})^{-1} = a \circ b\ \ \varepsilon\ H,$$

the closure (or binary) property of $\circ$ in H is established. The associative law for $\circ$ in H follows from the closure property. Thus, all the four postulates of a group are established making (H, $\circ$) a group and, therefore, a subgroup of (G, $\circ$). //

Replacing the general binary operation o by ordinary addition and multiplication there also follow the following conclusions from above theorem.

Corollary 2.1. A subset H of a group (G, +) forms a subgroup of (G, +) if there holds

$$\forall\ a, b\ \varepsilon\ \mathrm{H}\ \ \exists\ a - b\ \varepsilon\ \mathrm{H}.$$

Corollary 2.2. A subset H of a group (G, $\cdot$) forms a subgroup of (G, $\cdot$) if there holds

$$\forall\ a, b\ \varepsilon\ \mathrm{H}\ \exists\ ab^{-1}\ \varepsilon\ \mathrm{H}.$$

Definition 2.3. (*Centre of a group*) Let (G, $\cdot$) be a group and a subset C of G be defined by

$$\mathrm{C} = \{x : xa = ax\ \forall\ a\ \varepsilon\ \mathrm{G}\}, \tag{2.2}$$

then C is called the *centre* of the group (G, $\cdot$).

Theorem 2.4. The centre of a group forms a subgroup of the group.

Proof. As per definition of C we note that the identity element 1 (of G) also belongs to C for $a1 = 1a = a\ \varepsilon$ G. Next, $\forall\ x\ \varepsilon\ \mathrm{C} \Rightarrow x\ \varepsilon\ \mathrm{G} \Rightarrow x^{-1}$ ε G and for (2.2):

$x\,x^{-1} = x^{-1}x = 1$ (the identity element of C as well as G) $\Rightarrow x^{-1}\ \varepsilon$ C,

proving the unit character of every element of C.

Finally, $\forall\ x, y\ \varepsilon$ C there hold

$$ax = xa \quad \text{and} \quad ay = ya, \qquad \forall\ a\ \varepsilon\ \mathrm{G} \tag{2.3}$$

$\Rightarrow$

$$(x\,y)\,a = x\,(y\,a), \qquad \text{by associative law of group (G, }\cdot\text{)}$$

$$= x\,(a\,y), \qquad \text{by (2.3)}$$

$= (x\,a)\,y,$ by associative law of group (G, •)

$= (a\,x)\,y,$ again by (2.3)

$= a\,(x\,y),$ by associative law of group (G, •)

$\Rightarrow$

$x\,y\,\varepsilon$ C, by (2.2)

making the operation in C binary. The associative law of (G, •) is induced in C as well proving that (C, •) is also a group hence a subgroup of (G, •). //

Example 2.5. (*Reversal rule*) Given a group (G, •), there hold the rules:

$$\forall\, x, y\ \varepsilon\, \mathrm{G} \Rightarrow (x\,y)^{-1} = y^{-1}x^{-1}, \tag{2.4}$$

and

$$(x^{-1})^{-1} = x. \tag{2.5}$$

Solution. (i) In a group (G, ·) we have

$$\forall\, x, y\ \varepsilon\, \mathrm{G} \Rightarrow x\,y, x^{-1}, y^{-1}, y^{-1}x^{-1}\ \varepsilon\, \mathrm{G}.$$

Applying the associative law successively we, therefore, have

$$x\,y\,(y^{-1}x^{-1}) = x\,\{y\,(y^{-1}x^{-1})\}$$

$$= x\,\{(y\,y^{-1})\,x^{-1}\} = x\,(1\,x^{-1}) = x\,x^{-1} = 1,$$

and

$$(y^{-1}x^{-1})\,x\,y = y^{-1}\{x^{-1}(x\,y)\} =$$

$$y^{-1}\{(x^{-1}x)\,y\} = y^{-1}(1\,y) = y^{-1}y = 1.$$

Hence, the inverse of $x\,y$ is $y^{-1}x^{-1}$.

(ii) $\forall\, x\ \varepsilon\, \mathrm{G}\ \exists\, x^{-1}\ \varepsilon\, \mathrm{G}$ such that $x\,x^{-1} = x^{-1}x = 1$, showing that x and x^{-1} are inverses of each other: $x = (x^{-1})^{-1}$. //

Example 2.6. A group (G, •) with property $a^2 = e\ \forall\ a\ \varepsilon$ G is necessarily commutative.

Solution. The given property shows that every element of the group is its own inverse: $a = a^{-1}$, $b = b^{-1}\ \forall\ a, b\ \varepsilon$ G. Therefore, we have

$$a\,b = a^{-1}b^{-1} = (b\,a)^{-1}, \qquad \text{by (2.4)}$$

$$= b\,a, \qquad \text{as per hypothesis.}$$

This establishes the *abelian property* of the group. //

Example 2.7. For any positive integer m the set of residue classes w.r.t. modulo m:

$$\mathrm{I}_m = \{\bar{0}, \bar{1}, \bar{2}, \ldots, \overline{m-1}\} \qquad (2.6)$$

forms a group w.r.t. addition defined by

$$\bar{p} + \bar{q} = \bar{r}, \qquad (2.7)$$

where r is the remainder for division of the sum $p + q$ by m and all p, q, r vary from 0 to $m - 1$.

Solution. The addition is binary as per the definition (2.7). The class $\bar{0}$ acts as the identity element and inverse of any element $\bar{p}$ is $\overline{m-p}$ for $\bar{p} + \overline{m-p} = \bar{m} = \bar{0}$. Hence, ($\mathrm{I}_m$, +) is a group. In fact, it is also abelian as the sum in (2.7) is commutative. //

Example 2.8. The set of non-zero elements in (2.6):

$$\mathrm{I}_m{}^* = \{\bar{1}, \bar{2}, \ldots, \overline{m-1}\} \qquad (2.8)$$

also forms a group w.r.t. the multiplication defined by

$$\bar{p} \cdot \bar{q} = \bar{r}, \qquad (2.9)$$

where r is the remainder in the division of p . q by m and m is a prime number.

Solution. The multiplication so defined is binary in the set and 1 acts as the multiplicative identity. To check the inverse of any element p in the set we consider the problem in the following parts:

(i) When $m = 2$ (the first prime number), $\mathrm{I}_2{}^* = \{\bar{1}\}$ is a singleton with multiplicative identity. So, it forms a group.

(ii) Any prime number greater than 2 has to be an odd number, say 3. So, we consider the set of residue classes modulo 3: $I_3^* = \{\bar{1}, \bar{2}\}$, where $\bar{2}.\bar{2} = \bar{4} = \bar{1}$ (mod 3). Thus, either element in the set is unit.

(iii) Similarly, in $I_5^* = \{\bar{1}, \bar{2}, \bar{3}, \bar{4}\}$, $\bar{2}.\bar{3} = \bar{6} = \bar{1}$ (mod 5) $\Rightarrow \bar{2}$ and $\bar{3}$ are inverses of each other, while $\bar{4}.\bar{4} = \overline{16} = \bar{1}$ (mod 5) shows that $\bar{4}$ is the inverse of itself.

(iv) In $I_7^* = \{\bar{1}, \bar{2}, \bar{3}, \bar{4}, \bar{5}, \bar{6}\}$, $\bar{2}.\bar{4} = \bar{1}$, $\bar{3}.\bar{5} = \bar{1}$, $\bar{6}.\bar{6} = \bar{1}$ showing that every element in the set is unit.

(v) In $I_{11}^* = \{\bar{1}, \bar{2}, \bar{3}, \bar{4}, \bar{5}, \bar{6}, \bar{7}, \bar{8}, \bar{9}, \overline{10}\}$,

$$\bar{2}.\bar{6} = \bar{3}.\bar{4} = \bar{5}.\bar{9} = \bar{7}.\bar{8} = \overline{10}.\overline{10} = \bar{1} \text{ (mod 11)}$$

making every element in the set unit. //

§ 3. Some Special Types of Groups

3.1. Cyclic group

Let (G, ·) be a group and a be its any element. The integral powers of a (i.e. repeated multiples of a with itself) form a subset

$$\ell = \{a^p : p \in Z\} \tag{3.1}$$

of G. The set ℓ also forms a group w.r.t. multiplication, called the *cyclic group*. The identity element of (ℓ, ·) is the identity element of (G, ·) namely $1 = a^0$ (for $p = 0$) and the inverses of a^p are

$$a^{-p} = (a^{-1})^p = a^{-1}.a^{-1}\dots a^{-1} \text{ (repeated } p \text{ times).}$$

If for some finite value of p ($\neq 0$), a^p coincides with the identity element 1, the group is called a *finite cyclic group* and p its *order*, otherwise the order of the group is infinite. Also, a is called a *generator* of the group.

Example 3.1. The cube roots of unity form a cyclic group of order 3; and both ω and ω^2 can be taken as its generators.

Solution. The elements of the group can be generated by both ω and ω^2:

$$\{\omega, \omega^2, \omega^3 = 1\}, \qquad \text{or} \qquad \{\omega^2, (\omega^2)^2 = \omega, (\omega^2)^3 = 1\}. //$$

Example 3.2. The set of residue classes with respect to modulo 5 forms a cyclic group of order 5 w.r.t. addition of residue classes having every non–zero element as its generator.

Solution. The desired set is $\mathbb{I}_5 = \{\bar{0}, \bar{1}, \bar{2}, \bar{3}, \bar{4}\}$. Denoting its element $\bar{1}$ by a, we have

$$a + a = \bar{1} + \bar{1} = \bar{2}, \quad a + a + a = \bar{3}, \quad a + a + a + a = \bar{4},$$

$$a + a + a + a + a = \bar{0}.$$

Thus, every element of the set is an integral power (more precisely here multiple) of its generator $\bar{1}$. Similarly, beginning with $a = \bar{2}$, we have

$$a + a = \bar{4}, \ a + a + a = \bar{6} = \bar{1}, a + a + a + a = \bar{8} = \bar{3}$$

$$a + a + a + a + a = \overline{10} = \bar{0} \ (\text{mod } 5);$$

or, with $a = \bar{3}$,

$$a + a = \bar{1}, \ a + a + a = \bar{4}, a + a + a + a = \bar{2}, a + a + a + a + a = \bar{0}$$

and with $a = \bar{4}$,

$$a + a = \bar{3}, \ a + a + a = \bar{2}, a + a + a + a = \bar{1}, a + a + a + a + a = \bar{0}. //$$

3.2. Transformation group

Given a non-empty set G we consider one–to–one mappings of G onto itself. The set of these mappings, denoted by T (G), forms a group w.r.t. product of mappings. Such a group is called a *transformation group* of G. In case G is finite the transformation group T (G) is often called a *permutation group.*

Example 3.3. The permutations of the elements of the set S = {1, 2, 3} form the permutation group S_3 of degree 3 and order 3!.

Solution. We write all the 3! permutations of the set S:

$$S_3 = \{ P_1, P_2, P_3, P_4, P_5, P_6\}$$

Where

$$P_1 = \begin{pmatrix} 1 & 2 & 3 \\ 1 & 2 & 3 \end{pmatrix} = \textit{identity permutation}; \; P_2 = \begin{pmatrix} 1 & 2 & 3 \\ 2 & 1 & 3 \end{pmatrix} = (1, 2);$$

$$P_3 = \begin{pmatrix} 1 & 2 & 3 \\ 3 & 2 & 1 \end{pmatrix} = (1, 3); \; P_4 = \begin{pmatrix} 1 & 2 & 3 \\ 1 & 3 & 2 \end{pmatrix} = (2, 3);$$

$$P_5 = \begin{pmatrix} 1 & 2 & 3 \\ 2 & 3 & 1 \end{pmatrix} = \text{cycle } (1, 2, 3); \; P_6 = \begin{pmatrix} 1 & 2 & 3 \\ 3 & 1 & 2 \end{pmatrix} = \begin{pmatrix} 1 & 2 & 3 \\ 2 & 3 & 1 \end{pmatrix}\begin{pmatrix} 1 & 2 & 3 \\ 2 & 3 & 1 \end{pmatrix}$$

$$= (P_5)^2;$$

and note that

$$(P_2)^2 = \begin{pmatrix} 1 & 2 & 3 \\ 2 & 1 & 3 \end{pmatrix}\begin{pmatrix} 1 & 2 & 3 \\ 2 & 1 & 3 \end{pmatrix} = \begin{pmatrix} 1 & 2 & 3 \\ 1 & 2 & 3 \end{pmatrix} = P_1 \Rightarrow P_2 \text{ is } \textit{inverse} \text{ of itself};$$

$$P_2 P_3 = \begin{pmatrix} 1 & 2 & 3 \\ 2 & 1 & 3 \end{pmatrix} \begin{pmatrix} 1 & 2 & 3 \\ 3 & 2 & 1 \end{pmatrix} = \begin{pmatrix} 1 & 2 & 3 \\ 2 & 3 & 1 \end{pmatrix} = P_5;$$

$$P_2 P_4 = \begin{pmatrix} 1 & 2 & 3 \\ 2 & 1 & 3 \end{pmatrix} \begin{pmatrix} 1 & 2 & 3 \\ 1 & 3 & 2 \end{pmatrix} = \begin{pmatrix} 1 & 2 & 3 \\ 3 & 1 & 2 \end{pmatrix} = P_6;$$

$$P_2 P_5 = \begin{pmatrix} 1 & 2 & 3 \\ 2 & 1 & 3 \end{pmatrix} \begin{pmatrix} 1 & 2 & 3 \\ 2 & 3 & 1 \end{pmatrix} = \begin{pmatrix} 1 & 2 & 3 \\ 3 & 2 & 1 \end{pmatrix} = P_3;$$

$$P_2 P_6 = \begin{pmatrix} 1 & 2 & 3 \\ 2 & 1 & 3 \end{pmatrix}\begin{pmatrix} 1 & 2 & 3 \\ 3 & 1 & 2 \end{pmatrix} = \begin{pmatrix} 1 & 2 & 3 \\ 1 & 3 & 2 \end{pmatrix} = P_4;$$

$$P_3 P_2 = \begin{pmatrix} 1 & 2 & 3 \\ 3 & 2 & 1 \end{pmatrix} \begin{pmatrix} 1 & 2 & 3 \\ 2 & 1 & 3 \end{pmatrix} = \begin{pmatrix} 1 & 2 & 3 \\ 3 & 1 & 2 \end{pmatrix} = P_6 \neq P_2 P_3;$$

$$(P_3)^2 = \begin{pmatrix} 1 & 2 & 3 \\ 3 & 2 & 1 \end{pmatrix}\begin{pmatrix} 1 & 2 & 3 \\ 3 & 2 & 1 \end{pmatrix} = \begin{pmatrix} 1 & 2 & 3 \\ 1 & 2 & 3 \end{pmatrix} = P_1 \Rightarrow (P_3)^{-1} = P_3;$$

$$P_3 P_4 = \begin{pmatrix} 1 & 2 & 3 \\ 3 & 2 & 1 \end{pmatrix} \begin{pmatrix} 1 & 2 & 3 \\ 1 & 3 & 2 \end{pmatrix} = \begin{pmatrix} 1 & 2 & 3 \\ 2 & 3 & 1 \end{pmatrix} = P_5;$$

$$P_3 P_5 = \begin{pmatrix} 1 & 2 & 3 \\ 3 & 2 & 1 \end{pmatrix} \begin{pmatrix} 1 & 2 & 3 \\ 2 & 3 & 1 \end{pmatrix} = \begin{pmatrix} 1 & 2 & 3 \\ 1 & 3 & 2 \end{pmatrix} = P_4;$$

$$P_3\,P_6 = \begin{pmatrix}1 & 2 & 3\\ 3 & 2 & 1\end{pmatrix}\begin{pmatrix}1 & 2 & 3\\ 3 & 1 & 2\end{pmatrix} = \begin{pmatrix}1 & 2 & 3\\ 2 & 1 & 3\end{pmatrix} = P_2;$$

$$P_4\,P_2 = \begin{pmatrix}1 & 2 & 3\\ 1 & 3 & 2\end{pmatrix}\begin{pmatrix}1 & 2 & 3\\ 2 & 1 & 3\end{pmatrix} = \begin{pmatrix}1 & 2 & 3\\ 2 & 3 & 1\end{pmatrix} = P_5 \neq P_2\,P_4;$$

$$P_4\,P_3 = \begin{pmatrix}1 & 2 & 3\\ 1 & 3 & 2\end{pmatrix}\begin{pmatrix}1 & 2 & 3\\ 3 & 2 & 1\end{pmatrix} = \begin{pmatrix}1 & 2 & 3\\ 3 & 1 & 2\end{pmatrix} = P_6 \neq P_3\,P_4;$$

$$(P_4)^2 = \begin{pmatrix}1 & 2 & 3\\ 1 & 3 & 2\end{pmatrix}\begin{pmatrix}1 & 2 & 3\\ 1 & 3 & 2\end{pmatrix} = \begin{pmatrix}1 & 2 & 3\\ 1 & 2 & 3\end{pmatrix} = P_1 \Rightarrow (P_4)^{-1} = P_4;$$

$$P_4\,P_5 = \begin{pmatrix}1 & 2 & 3\\ 1 & 3 & 2\end{pmatrix}\begin{pmatrix}1 & 2 & 3\\ 2 & 3 & 1\end{pmatrix} = \begin{pmatrix}1 & 2 & 3\\ 2 & 1 & 3\end{pmatrix} = P_2;$$

$$P_4\,P_6 = \begin{pmatrix}1 & 2 & 3\\ 1 & 3 & 2\end{pmatrix}\begin{pmatrix}1 & 2 & 3\\ 3 & 1 & 2\end{pmatrix} = \begin{pmatrix}1 & 2 & 3\\ 3 & 2 & 1\end{pmatrix} = P_3;$$

$$P_5\,P_2 = \begin{pmatrix}1 & 2 & 3\\ 2 & 3 & 1\end{pmatrix}\begin{pmatrix}1 & 2 & 3\\ 2 & 1 & 3\end{pmatrix} = \begin{pmatrix}1 & 2 & 3\\ 1 & 3 & 2\end{pmatrix} = P_4 \neq P_2\,P_5;$$

$$P_5\,P_3 = \begin{pmatrix}1 & 2 & 3\\ 2 & 3 & 1\end{pmatrix}\begin{pmatrix}1 & 2 & 3\\ 3 & 2 & 1\end{pmatrix} = \begin{pmatrix}1 & 2 & 3\\ 2 & 1 & 3\end{pmatrix} = P_2 \neq P_3\,P_5;$$

$$P_5\,P_4 = \begin{pmatrix}1 & 2 & 3\\ 2 & 3 & 1\end{pmatrix}\begin{pmatrix}1 & 2 & 3\\ 1 & 3 & 2\end{pmatrix} = \begin{pmatrix}1 & 2 & 3\\ 3 & 2 & 1\end{pmatrix} = P_3 \neq P_4\,P_5;$$

$$(P_5)^2 = \begin{pmatrix}1 & 2 & 3\\ 2 & 3 & 1\end{pmatrix}\begin{pmatrix}1 & 2 & 3\\ 2 & 3 & 1\end{pmatrix} = \begin{pmatrix}1 & 2 & 3\\ 3 & 1 & 2\end{pmatrix} = P_6;$$

$$P_5\,P_6 = \begin{pmatrix}1 & 2 & 3\\ 2 & 3 & 1\end{pmatrix}\begin{pmatrix}1 & 2 & 3\\ 3 & 1 & 2\end{pmatrix} = \begin{pmatrix}1 & 2 & 3\\ 1 & 2 & 3\end{pmatrix} = P_1$$

$\Rightarrow$ P_5, P_6 are inverses of each other;

$$P_6\,P_2 = \begin{pmatrix}1 & 2 & 3\\ 3 & 1 & 2\end{pmatrix}\begin{pmatrix}1 & 2 & 3\\ 2 & 1 & 3\end{pmatrix} = \begin{pmatrix}1 & 2 & 3\\ 3 & 2 & 1\end{pmatrix} = P_3 \neq P_2\,P_6;$$

$$P_6\,P_3 = \begin{pmatrix}1 & 2 & 3\\ 3 & 1 & 2\end{pmatrix}\begin{pmatrix}1 & 2 & 3\\ 3 & 2 & 1\end{pmatrix} = \begin{pmatrix}1 & 2 & 3\\ 1 & 3 & 2\end{pmatrix} = P_4 \neq P_3\,P_6;$$

$$P_6\,P_4 = \begin{pmatrix}1 & 2 & 3\\ 3 & 1 & 2\end{pmatrix}\begin{pmatrix}1 & 2 & 3\\ 1 & 3 & 2\end{pmatrix} = \begin{pmatrix}1 & 2 & 3\\ 2 & 1 & 3\end{pmatrix} = P_2 \neq P_4\,P_6;$$

$$P_6\,P_5 = \begin{pmatrix}1 & 2 & 3\\ 3 & 1 & 2\end{pmatrix}\begin{pmatrix}1 & 2 & 3\\ 2 & 3 & 1\end{pmatrix} = \begin{pmatrix}1 & 2 & 3\\ 1 & 2 & 3\end{pmatrix} = P_1;$$

$$(P_6)^2 = \begin{pmatrix}1 & 2 & 3\\ 3 & 1 & 2\end{pmatrix}\begin{pmatrix}1 & 2 & 3\\ 3 & 1 & 2\end{pmatrix} = \begin{pmatrix}1 & 2 & 3\\ 2 & 3 & 1\end{pmatrix} = P_5\,.$$

Thus, the permutation multiplication is binary in the set, P_1 acts as the identity permutation, P_2 to P_4 are inverses of themselves, and P_5, P_6 are inverses of each other.

Hence, the set S_3 forms a group (however, not commutative) w.r.t. permutation multiplication. //

Note 3.1. The permutation group in above example is called a *symmetric group of order* 3!.

Similarly, the permutation group S_n of the set $\{1, 2, \ldots, n\}$ of n elements will be a symmetric group of order n!.

IV. Vector Algebra

Fig. 1.1

§ 1. Physical Quantities

In our discussion, we shall mainly encounter with two types of quantities: scalars and vectors. Those having only magnitude are called *scalars*. For instance, area, density, length, mass, speed, volume, numbers, etc., are all scalars. On the other hand, the physical quantities equipped with both magnitude and direction are called *vectors*. Acceleration, displacement, electric field, force, magnetic field, velocity, weight, etc., are vectors.

Let us consider a line OP of length, say a. The directed line $\overrightarrow{OP}$ represents a vector (from O to P). It may be denoted by a bold face letter, say **a**, while a scalar will be written in *italic* letter. The length a of the line OP is called the *magnitude* of the vector **a**. In symbols we denote it by $|\,\mathbf{a}\,| = a$.

Note 1.1. A vector with zero magnitude is called a *zero* or *null vector*. It is denoted by **0**. Its direction is *indeterminate*.

1.1. Addition of vectors

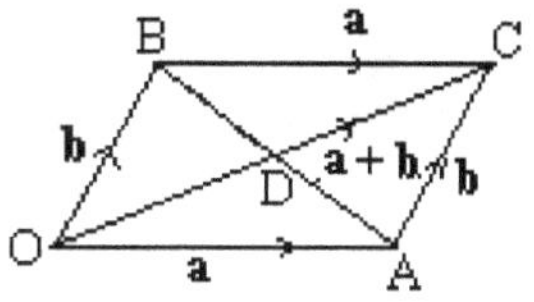

Fig. 1.2

Let $\overrightarrow{OA}$ = **a** and $\overrightarrow{OB}$ = **b** be two vectors acting at a point O. Complete the parallelogram OACB and draw its diagonals OC and BA. The vector along the diagonal $\overrightarrow{OC}$ is defined as the *vector sum* of the vectors $\overrightarrow{OA}$ and $\overrightarrow{OB}$. The opposite sides OB and AC being equal and parallel and the vectors along them are in same sense of direction. So they are equal. We, therefore, write it as

$$\overrightarrow{OC} = \overrightarrow{OA} + \overrightarrow{AC} = \overrightarrow{OA} + \overrightarrow{OB} = \mathbf{a} + \mathbf{b}. \qquad (1.1)$$

Similarly, $\overrightarrow{OA} = \overrightarrow{BC}$ = **a**. The vector along the other diagonal BA is defined as the *difference* of two vectors $\overrightarrow{OA}$ and $\overrightarrow{OB}$:

$$\overrightarrow{BA} = \overrightarrow{OA} - \overrightarrow{OB} = \mathbf{a} - \mathbf{b}. \quad (1.2)$$

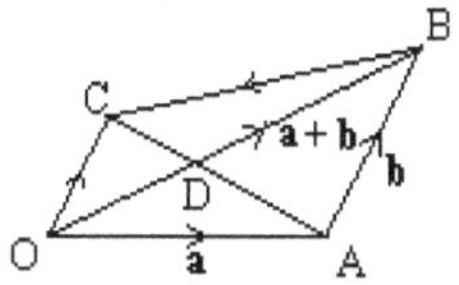

Fig. 1.3

Theorem 1.1. The vector sum is commutative and associative:

$$\mathbf{a} + \mathbf{b} = \mathbf{b} + \mathbf{a}, \qquad (1.3)$$

$$\mathbf{a} + (\mathbf{b} + \mathbf{c}) = (\mathbf{a} + \mathbf{b}) + \mathbf{c} \qquad (1.4)$$

Proof. (i) Cf. Fig.1.2. We have

$$\overrightarrow{OC} = \overrightarrow{OB} + \overrightarrow{BC} = \overrightarrow{OB} + \overrightarrow{OA} = \mathbf{b} + \mathbf{a},$$

which, in association with (1.1), establishes (1.3).

(ii) In Fig.1.3, LHS of Eq. (1.4) = $\mathbf{a} + \overrightarrow{AC} = \overrightarrow{OC}$. Also, its

$$\text{RHS} = \overrightarrow{OB} + \mathbf{c} = \overrightarrow{OC}. \ //$$

Corollary 1.1. For every vector **a**, there hold $\mathbf{a} + \mathbf{0} = \mathbf{0} + \mathbf{a} = \mathbf{a}$.

Example 1.1. Three vectors **a**, **b**, **c** act along the consecutive sides of a triangle. Show that their vector sum vanishes. Generalize the result for any closed polygon.

Solution. Let the vectors form a triangle ABC where $\overrightarrow{AB} = \mathbf{a}$, $\overrightarrow{BC} = \mathbf{b}$ and $\overrightarrow{CA} = \mathbf{c}$. By definition of vector sum,

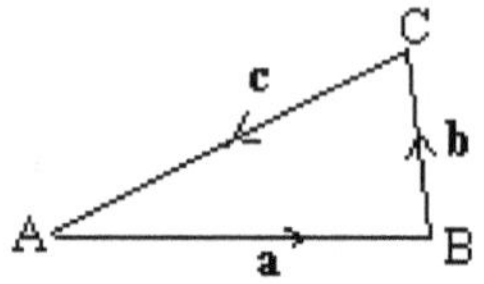

Fig. 1.4

$$\overrightarrow{AB} + \overrightarrow{BC} = \overrightarrow{AC}, \quad \text{i.e.} \quad \mathbf{a} + \mathbf{b} = -\mathbf{c}$$

$$\Rightarrow \quad \mathbf{a} + \mathbf{b} + \mathbf{c} = \mathbf{c} + (-\mathbf{c}) = \mathbf{0}.$$

Generalization can be similarly proved.//

Example 1.2. ABCD is a parallelogram with G as the point of intersection of its diagonals. For any point O prove that

$$\overrightarrow{OA} + \overrightarrow{OB} + \overrightarrow{OC} + \overrightarrow{OD} = 4\overrightarrow{OG}.$$

Solution. The vector $\overrightarrow{OG}$ can be expressed as the sum of vectors:

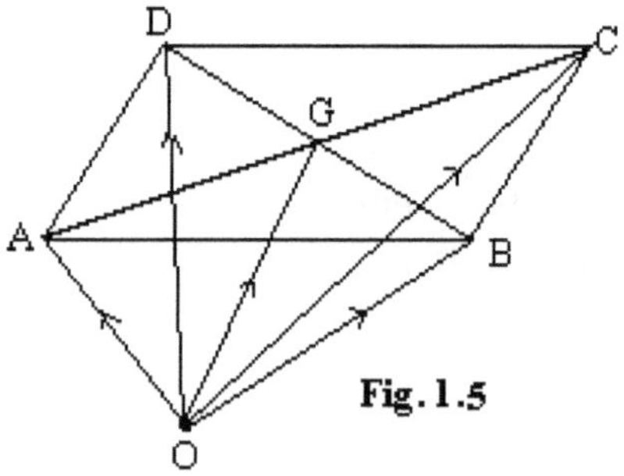

Fig. 1.5

$$\overrightarrow{OG} = \overrightarrow{OA} + \overrightarrow{AG},$$

$\overrightarrow{OG} = \overrightarrow{OB} + \overrightarrow{BG}$, $\overrightarrow{OG} = \overrightarrow{OC} + \overrightarrow{CG}$

and

$$\overrightarrow{OG} = \overrightarrow{OD} + \overrightarrow{DG}.$$

Adding these vectors, we get

$$4\,\overrightarrow{OG} = (\overrightarrow{OA} + \overrightarrow{OB} + \overrightarrow{OC} + \overrightarrow{OD}) + (\overrightarrow{AG} + \overrightarrow{BG} + \overrightarrow{CG} + \overrightarrow{DG}).$$

The diagonals of a parallelogram bisect each other. Thus, G being the midpoint of both diagonals, we have $\overrightarrow{AG} + \overrightarrow{CG} = \overrightarrow{BG} + \overrightarrow{DG} = \mathbf{0}$. This reduces above result to the desired form.

§ 2. Scalar Multiplication of a Vector

Let x be a scalar and $\mathbf{a}$ some vector then $x\mathbf{a}$ is also a vector with magnitude $|\,x\,|$ times that of $\mathbf{a}$. The direction of $x\mathbf{a}$ will be along (or opposite to) $\mathbf{a}$ if x is positive (or negative).

Definition 2.1. Two vectors **a** and **b** are called *collinear* iff one of them is some scalar multiple of the other:

$$\mathbf{a} = t\,\mathbf{b}, \tag{2.1}$$

where $t \neq 0$ is some scalar. The magnitudes of collinear vectors are proportional. If, in addition, their magnitudes are same above vectors are called *equal* (or *opposite*) when $t = 1$ (or -1).

Theorem 2.1. Scalar multiplication of vector(s) satisfies the following laws:

(*Associative law*) $$x\,(y\,\mathbf{a}) = (x\,y)\,\mathbf{a}, \tag{2.2}$$

(*Distributive laws*)

$$(x + y)\,\mathbf{a} = x\,\mathbf{a} + y\,\mathbf{a},\ \ x\,(\mathbf{a} + \mathbf{b}) = x\,\mathbf{a} + x\,\mathbf{b}. \tag{2.3}$$

Proof being simple is left for the reader. //

Definition 2.2. A vector with unit magnitude is called a *unit vector*.

The unit vector along any non–null vector **a** is obtained by multiplying the vector by the reciprocal of its magnitude:

$$\hat{\mathbf{a}} \equiv \mathbf{a}\,/\,|\,\mathbf{a}\,|\,. \tag{2.4}$$

Example 2.1. Prove the following results geometrically:

$$\mathbf{a} = (\mathbf{a} + \mathbf{b})/2 + (\mathbf{a} - \mathbf{b})/2, \tag{2.5}$$

$$\mathbf{b} = (\mathbf{a} + \mathbf{b})/2 - (\mathbf{a} - \mathbf{b})/2. \tag{2.6}$$

Solution. Cf. Fig. 1.2. The diagonals OC and BA of the parallelogram bisect each other in the point D. Therefore,

$$\overrightarrow{OD} = (1/2)\overrightarrow{OC} = (\mathbf{a} + \mathbf{b})/2 \quad \text{and} \quad \overrightarrow{DA} = (1/2)\overrightarrow{BA} = (\mathbf{a} - \mathbf{b})/2.$$

Their sum determines the vector $\overrightarrow{OA} = \mathbf{a}$. This establishes (2.5).

Next, in the Δ OBD: $\overrightarrow{OB} = \overrightarrow{OD} - \overrightarrow{BD} = (1/2)\,\overrightarrow{OC} - (1/2)\overrightarrow{BA}$ giving (2.6). //

Example 2.2. Show that the sum of vectors along the medians (directed from the vertices) of a triangle vanishes.

Solution. Let AD, BE and CF be the medians of the triangle ABC. The vectors $\overrightarrow{BD}$ and $\overrightarrow{BC}$ are in same sense of direction and the former is of half magnitude to that of the latter. Hence,

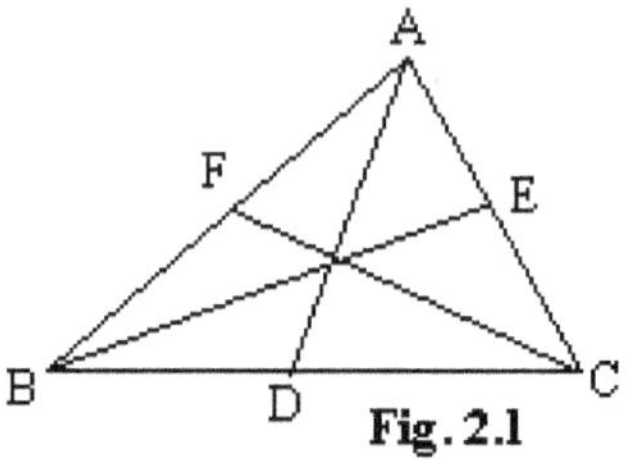

Fig. 2.1

$$\overrightarrow{BD} = (1/2)\overrightarrow{BC} \Rightarrow \overrightarrow{AD} = \overrightarrow{AB} + \overrightarrow{BD} = \overrightarrow{AB} + (1/2)\overrightarrow{BC}.$$

Similarly,

$$\overrightarrow{BE} = \overrightarrow{BC} + (1/2)\overrightarrow{CA} \quad \text{and} \quad \overrightarrow{CF} = \overrightarrow{CA} + (1/2)\overrightarrow{AB}.$$

Adding these vectors, by Example 1.1, we get

$$\overrightarrow{AD} + \overrightarrow{BE} + \overrightarrow{CF} = (3/2)(\overrightarrow{BC} + \overrightarrow{CA} + \overrightarrow{AB}) = \mathbf{0}.$$

§ 3. Components of a Vector Along the Coordinate Axes

Let P(x, y, z) be any point in the space E_3 with position vector $\overrightarrow{OP} = \mathbf{r}$ with respect to some origin O and three rectangular coordinate axes Ox, Oy, Oz through O. Let us draw the perpendicular PL from P to the plane xOy. Perpendiculars from L to the coordinate axes Ox and Oy are also drawn. If $\hat{\mathbf{i}}$, $\hat{\mathbf{j}}$, $\hat{\mathbf{k}}$ be the unit vectors along the respective coordinate axes then,

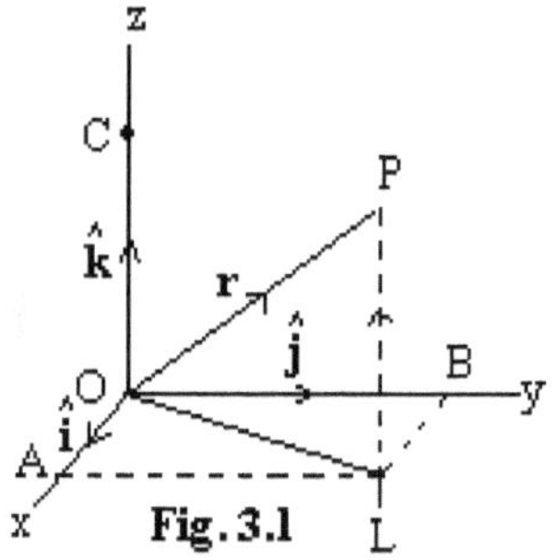

Fig. 3.1

$$\overrightarrow{OA} = (\text{OA})\,\hat{\mathbf{i}} = x\,\hat{\mathbf{i}}, \qquad \overrightarrow{OB} = (\text{OB})\,\hat{\mathbf{j}} = y\,\hat{\mathbf{j}}.$$

Similarly, $\overrightarrow{OC} = \overrightarrow{LP} = z\hat{\mathbf{k}}$, where C is taken on O$z$ line at a distance equal to z from O. By the definition of vector addition, we have

$$\overrightarrow{OA} + \overrightarrow{AL} = \overrightarrow{OL} \quad \text{and} \quad \overrightarrow{OL} + \overrightarrow{LP} = \overrightarrow{OP}.$$

Therefore,

$$\overrightarrow{OP} \equiv \mathbf{r} = \overrightarrow{OA} + \overrightarrow{AL} + \overrightarrow{LP} = \overrightarrow{OA} + \overrightarrow{OB} + \overrightarrow{OC}$$
$$= x\hat{\mathbf{i}} + y\hat{\mathbf{j}} + z\hat{\mathbf{k}} \equiv (x, y, z)\,. \tag{3.1}$$

Definition 3.1. The numbers x, y, z are called the *components* of the vector $\overrightarrow{OP}$ along the rectangular coordinate axes Ox, Oy, Oz respectively. They also determine the coordinates of the point P. In particular,

$$\left.\begin{array}{l} \mathbf{0} = (0, 0, 0) = 0\,\hat{\mathbf{i}} + 0\,\hat{\mathbf{j}} + 0\,\hat{\mathbf{k}}\,,\ \hat{\mathbf{i}} = (1, 0, 0) = 1\hat{\mathbf{i}} + 0\,\hat{\mathbf{j}} + 0\,\hat{\mathbf{k}}\,, \\ \hat{\mathbf{j}} = (0, 1, 0) = 0\,\hat{\mathbf{i}} + 1\hat{\mathbf{j}} + 0\,\hat{\mathbf{k}}\,,\ \ \hat{\mathbf{k}} = (0, 0, 1) = 0\,\hat{\mathbf{i}} + 0\,\hat{\mathbf{j}} + 1\,\hat{\mathbf{k}} \end{array}\right\} . \tag{3.2}$$

Theorem 3.1. The vectors

$$\mathbf{a} = (a_1, a_2, a_3) \text{ and } \mathbf{b} = (b_1, b_2, b_3) \tag{3.3}$$

are equal if $a_i = b_i$, $i = 1, 2, 3$.

Proof. The vectors

$$\mathbf{a} = a_1\hat{\mathbf{i}} + a_2\hat{\mathbf{j}} + a_3\hat{\mathbf{k}} \text{ and } \mathbf{b} = b_1\hat{\mathbf{i}} + b_2\hat{\mathbf{j}} + b_3\hat{\mathbf{k}} \tag{3.4}$$

are equal if

$$(a_1 - b_1)\,\hat{\mathbf{i}} + (a_2 - b_2)\,\hat{\mathbf{j}} + (a_3 - b_3)\,\hat{\mathbf{k}} = \mathbf{0} = (0, 0, 0)$$

$\Rightarrow$

$$a_1 - b_1 = a_2 - b_2 = a_3 - b_3 = 0 \Rightarrow a_1 = b_1,\ a_2 = b_2,\ a_3 = b_3.$$

Conversely,

$$a_1 = b_1,\ a_2 = b_2,\ a_3 = b_3 \Rightarrow a_1\hat{\mathbf{i}} = b_1\hat{\mathbf{i}},\ a_2\hat{\mathbf{j}} = b_2\hat{\mathbf{j}},\ a_3\hat{k} = b_3\hat{\mathbf{k}}\,.$$

Their addition determines

$$a_1\hat{\mathbf{i}} + a_2\hat{\mathbf{j}} + a_3\hat{\mathbf{k}} = b_1\hat{\mathbf{i}} + b_2\hat{\mathbf{j}} + b_3\hat{\mathbf{k}}, \qquad \text{i.e. } \mathbf{a} = \mathbf{b}.$$

Hence, the condition is sufficient as well.

Theorem 3.2. Given a scalar x and a vector **a**, given by (3.3), their scalar multiplication is a vector

$$x\,\mathbf{a} = (xa_1, xa_2, xa_3)\,. \tag{3.5}$$

Proof. Multiplying the vector **a**, as given by (3.4), by the scalar x, we get

$$x\mathbf{a} = x\,(a_1\hat{\mathbf{i}} + a_2\hat{\mathbf{j}} + a_3\hat{\mathbf{k}}) = (xa_1)\,\hat{\mathbf{i}} + (xa_2)\,\hat{\mathbf{j}} + (xa_3)\,\hat{\mathbf{k}} = (xa_1, xa_2, xa_3).$$

Corollary 3.1. The vector $-\,\mathbf{a} = (-\,a_1, -\,a_2, -\,a_3)$ is the negative of the vector **a** given by (3.3).

Definition 3.2. The vector sum of two vectors in (3.3) is defined by

$$\mathbf{a} + \mathbf{b} = (a_1 + b_1, a_2 + b_2, a_3 + b_3), \tag{3.6}$$

whereas the scalar multiple of a vector **a** in (3.3) is defined by (3.5).

§ 4. Products of Vectors

4.1. Dot product of two vectors

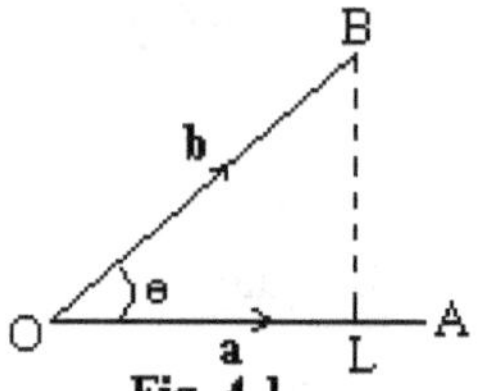

Fig. 4.1

Let **a** and **b** be two non–null vectors inclined at an angle θ (measured from **a** to **b**). Their *dot product* **a . b** is defined by

$$\mathbf{a} \cdot \mathbf{b} = |\,\mathbf{a}\,|\,|\,\mathbf{b}\,|\cos\theta. \tag{4.1}$$

Being a scalar, it is also called the *scalar* (or *inner*) *product* of two vectors. For $\cos(-\,\theta) = \cos\theta$,

$$\mathbf{b} \cdot \mathbf{a} = |\,\mathbf{b}\,|\,|\,\mathbf{a}\,|\cos(-\,\theta) = \mathbf{a} \cdot \mathbf{b}\,. \tag{4.2}$$

Thus, the dot product is *commutative*. Replacing **b** by **a** in (4.1), there follows

$$\mathbf{a} \cdot \mathbf{a} = |\,\mathbf{a}\,|\,|\,\mathbf{a}\,|\cos 0 = |\,\mathbf{a}\,|^2 \;\Rightarrow\; |\,\mathbf{a}\,| = +\sqrt{(\mathbf{a} \cdot \mathbf{a})}\,. \tag{4.3}$$

Also, the unit vectors $\hat{\mathbf{i}}$, $\hat{\mathbf{j}}$, $\hat{\mathbf{k}}$ acting along the coordinate axes satisfy

$$\hat{\mathbf{i}} \cdot \hat{\mathbf{i}} = \hat{\mathbf{j}} \cdot \hat{\mathbf{j}} = \hat{\mathbf{k}} \cdot \hat{\mathbf{k}} = 1, \quad \hat{\mathbf{i}} \cdot \hat{\mathbf{j}} = \hat{\mathbf{j}} \cdot \hat{\mathbf{i}} = \hat{\mathbf{j}} \cdot \hat{\mathbf{k}} = \hat{\mathbf{k}} \cdot \hat{\mathbf{j}} = \hat{\mathbf{k}} \cdot \hat{\mathbf{i}} = \hat{\mathbf{i}} \cdot \hat{\mathbf{k}} = 0. \tag{4.4}$$

Accordingly, the dot product of the vector in (3.1) with itself yields

$$\mathbf{r} \cdot \mathbf{r} = |\,\mathbf{r}\,|^2 = (x\hat{\mathbf{i}} + y\hat{\mathbf{j}} + z\hat{\mathbf{k}}) \cdot (x\hat{\mathbf{i}} + y\hat{\mathbf{j}} + z\hat{\mathbf{k}}) = x^2 + y^2 + z^2 \tag{4.5}$$

$\Rightarrow$

$$|\,\mathbf{r}\,| = \sqrt{(x^2 + y^2 + z^2)} = r \text{ (say)}. \tag{4.6}$$

The unit vector along **r** is

$$\hat{\mathbf{r}} = \mathbf{r} / |\mathbf{r}| = (x/r)\,\hat{\mathbf{i}} + (y/r)\,\hat{\mathbf{j}} + (z/r)\,\hat{\mathbf{k}}\,. \tag{4.7}$$

Theorem 4.1. Two non–null vectors in (3.3) are orthogonal to each other if

$$\mathbf{a}\,.\,\mathbf{b} = 0. \tag{4.8}$$

Proof. For mutually perpendicular vectors their dot product (4.1) becomes

$$\mathbf{a}\,.\,\mathbf{b} = |\,\mathbf{a}\,|\,|\,\mathbf{b}\,| \cos 90 = 0.$$

This proves the necessary part. Conversely, when the condition (4.8) holds there follows $\cos\theta = 0$, as the vectors being non–null. Hence, $\theta = 90°$. //

Theorem 4.2. The dot product of two vectors in (3.4) is

$$\mathbf{a}\,.\,\mathbf{b} = a_1\, b_1 + a_2\, b_2 + a_3\, b_3\,. \tag{4.9}$$

Proof. Forming the dot product of two vectors and putting from (4.4) we get the result. //

Corollary 4.1. The magnitude of the vector **a**, given by (3.3), is

$$|\,\mathbf{a}\,| = \sqrt{(\,a_1^2 + a_2^2 + a_3^2)}\,. \tag{4.10}$$

Proof. Replacing **b** by **a** in (4.9) and using (4.3) we obtain the desired result. //

Theorem 4.3. Angle between two vectors **a** and **b** can be measured by

$$\cos\theta = \mathbf{a}\,.\,\mathbf{b} / |\,\mathbf{a}\,|\,|\,\mathbf{b}\,|$$

$$= (a_1b_1 + a_2b_2 + a_3b_3) / \sqrt{(a_1^2 + a_2^2 + a_3^2)}\,\sqrt{(b_1^2 + b_2^2 + b_3^2)}\,.$$

Proof. Analogous to (4.10) we also have

$$|\,\mathbf{b}\,| = \sqrt{(\,b_1^2 + b_2^2 + b_3^2)}\,. \tag{4.12}$$

Putting from (4.9), (4.10) and (4.12) in (4.1), we get the result. //

Corollary 4.2. The vectors given by (3.3) are orthogonal to each other if

$$a_1 b_1 + a_2 b_2 + a_3 b_3 = 0. \tag{4.13}$$

Proof. The result follows immediately from (4.8) and (4.9). //

Definition 4.1. Mutually orthogonal unit vectors are called *orthonormal* vectors.

The unit vectors $\hat{\mathbf{i}}$, $\hat{\mathbf{j}}$, $\hat{\mathbf{k}}$ acting along the respective rectangular coordinate axes Ox, Oy and Oz are orthonormal vectors.

Definition 4.2. Let BL be drawn perpendicular from B to $\overrightarrow{OA} = \mathbf{a}$ in Fig. 4.1. The length OL given by

$$\text{OL} = (\text{OB}) \cos\theta = |\mathbf{b}| \cos\theta = |\mathbf{b}|\,|\mathbf{a}|\,(\cos\theta)\,/\,|\mathbf{a}|$$

$$= \mathbf{b}\,.\,\mathbf{a}\,/\,a = \overrightarrow{OB}\,.\,\overrightarrow{OA}\,/\,\text{OA} \tag{4.14}$$

is called the *resolved part of vector* **b** *along the vector* **a**.

Theorem 4.4. The dot product is distributive over vector addition:

$$\mathbf{a}\,.\,(\mathbf{b} + \mathbf{c}) = \mathbf{a}\,.\,\mathbf{b} + \mathbf{a}\,.\,\mathbf{c}. \tag{4.15}$$

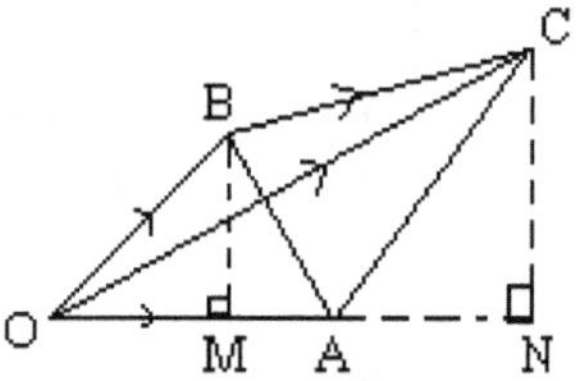

Fig. 4.2

Proof. Let **a**, **b**, **c** be the position vectors of three points A, B, C with respect to some origin O. Join OC and draw the perpendiculars BM and CN to $\overrightarrow{OA} = \mathbf{a}$, so that OM, MN and ON are the resolved parts of vectors $\overrightarrow{OB} = \mathbf{b}$, $\overrightarrow{BC} = \mathbf{c}$ and $\overrightarrow{OC} = \overrightarrow{OB} + \overrightarrow{BC} = \mathbf{b} + \mathbf{c}$. Therefore, the LHS of Eq. (4.15) is

$$a\,\{\text{resolved part of } (\mathbf{b} + \mathbf{c}) \text{ along } \mathbf{a}\} = a\,(\text{ON})$$

$$= a\,(\text{OM} + \text{MN}) = a\,(\text{OM}) + a\,(\text{MN})$$

$$= a\,(\text{resolved part of } \mathbf{b} \text{ along } \mathbf{a}) + a\,(\text{resolved part of } \mathbf{c} \text{ along } \mathbf{a})$$

$$= \mathbf{a}\,.\,\mathbf{b} + \mathbf{a}\,.\,\mathbf{c} = \text{RHS of Eq. (4.15)}.$$

4.2. Cross product of two vectors

The product of two vectors **a** and **b** defined by

$$\mathbf{a} \times \mathbf{b} \equiv |\mathbf{a}|\,|\mathbf{b}|\,(\sin\theta)\,\hat{\mathbf{u}}, \qquad (4.16)$$

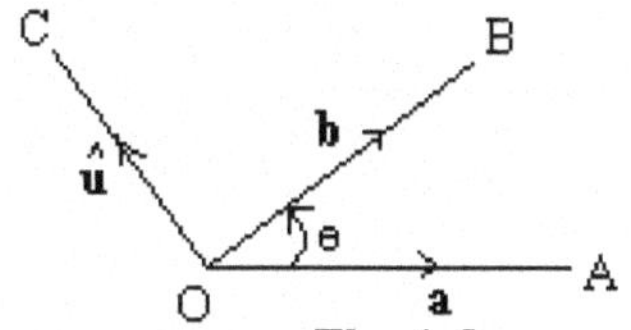

Fig. 4.3

where $\hat{\mathbf{u}}$ is the unit vector orthogonal to both **a** and **b** such that **a**, **b** and $\hat{\mathbf{u}}$ form a right-handed system. The cross product being a vector is also called the *vector* (or *outer*) *product.* In contrary to (4.2), the cross product of two vectors is *anti–commutative* (i.e. skew–symmetric):

$$\mathbf{b} \times \mathbf{a} = |\mathbf{b}|\,|\mathbf{a}|\,(\sin\theta)\,(-\hat{\mathbf{u}}) = -|\mathbf{b}|\,|\mathbf{a}|\,(\sin\theta)\,\hat{\mathbf{u}} = -(\mathbf{a} \times \mathbf{b}). \quad (4.17)$$

Analogous to the relations (4.3) and (4.4) there hold:

$$\left.\begin{array}{l} \mathbf{a} \times \mathbf{a} = |\mathbf{a}|^2 (\sin 0)\,\hat{\mathbf{u}} = \mathbf{0}, \quad \hat{\mathbf{i}} \times \hat{\mathbf{i}} = \hat{\mathbf{j}} \times \hat{\mathbf{j}} = \hat{\mathbf{k}} \times \hat{\mathbf{k}} = \mathbf{0}, \\ \hat{\mathbf{i}} \times \hat{\mathbf{j}} = -\hat{\mathbf{j}} \times \hat{\mathbf{i}} = \hat{\mathbf{k}},\ \hat{\mathbf{j}} \times \hat{\mathbf{k}} = -\hat{\mathbf{k}} \times \hat{\mathbf{j}} = \hat{\mathbf{i}},\ \hat{\mathbf{k}} \times \hat{\mathbf{i}} = -\hat{\mathbf{i}} \times \hat{\mathbf{k}} = \hat{\mathbf{j}}. \end{array}\right\} \quad (4.18)$$

Note 4.1. The unit vectors $\hat{\mathbf{i}}$, $\hat{\mathbf{j}}$, $\hat{\mathbf{k}}$ acting along the respective coordinate axes form a right-handed system, and so do the triads: $\hat{\mathbf{j}}$, $\hat{\mathbf{k}}$, $\hat{\mathbf{i}}$ and $\hat{\mathbf{k}}$, $\hat{\mathbf{i}}$, $\hat{\mathbf{j}}$. An interchange of any two vectors in these triads makes them left-handed. Thus, $\hat{\mathbf{j}}$, $\hat{\mathbf{i}}$, $\hat{\mathbf{k}}$; $\hat{\mathbf{k}}$, $\hat{\mathbf{j}}$, $\hat{\mathbf{i}}$; and $\hat{\mathbf{i}}$, $\hat{\mathbf{k}}$, $\hat{\mathbf{j}}$ are in the left-handed system. On the other hand, $\hat{\mathbf{j}}$, $\hat{\mathbf{i}}$, $-\hat{\mathbf{k}}$; $\hat{\mathbf{k}}$, $\hat{\mathbf{j}}$, $-\hat{\mathbf{i}}$; and $\hat{\mathbf{i}}$, $\hat{\mathbf{k}}$, $-\hat{\mathbf{j}}$ again form right-handed systems.

Theorem 4.5. The cross product of two vectors **a** and **b**, given in (3.3), is

$$\mathbf{a} \times \mathbf{b} = \begin{vmatrix} \hat{\mathbf{i}} & \hat{\mathbf{j}} & \hat{\mathbf{k}} \\ a_1 & a_2 & a_3 \\ b_1 & b_2 & b_3 \end{vmatrix}$$

$$= (a_2 b_3 - a_3 b_2)\,\hat{\mathbf{i}} + (a_3 b_1 - a_1 b_3)\,\hat{\mathbf{j}} + (a_1 b_2 - a_2 b_1)\,\hat{\mathbf{k}}\,. \quad (4.19)$$

Proof. $\mathbf{a} \times \mathbf{b} = (a_1\hat{\mathbf{i}} + a_2\hat{\mathbf{j}} + a_3\hat{\mathbf{k}}) \times (b_1\hat{\mathbf{i}} + b_2\hat{\mathbf{j}} + b_3\hat{\mathbf{k}})$

$$= (a_1 b_2 - a_2 b_1)(\hat{\mathbf{i}} \times \hat{\mathbf{j}}) + (a_2 b_3 - a_3 b_2)(\hat{\mathbf{j}} \times \hat{\mathbf{k}}) + (a_3 b_1 - a_1 b_3)(\hat{\mathbf{k}} \times \hat{\mathbf{i}})$$

$= (a_1 b_2 - a_2 b_1)\,\hat{\mathbf{k}} + (a_2 b_3 - a_3 b_2)\,\hat{\mathbf{\imath}} + (a_3 b_1 - a_1 b_3)\,\hat{\mathbf{\jmath}},$ by (4.18). //

Theorem 4.6. Angle between two vectors can also be measured by

$$\sin^2\theta = \{(a_1 b_2 - a_2 b_1)^2 + (a_2 b_3 - a_3 b_2)^2 + (a_3 b_1 - a_1 b_3)^2\}$$

$$\div\ (a_1^2 + a_2^2 + a_3^2)\,(b_1^2 + b_2^2 + b_3^2). \tag{4.20}$$

Proof. Forming the dot product of (4.19) with itself, using (4.16) and putting from (4.4) we get

$$|\,\mathbf{a}\,|^2\,|\,\mathbf{b}\,|^2\,(\sin^2\theta)\,(\hat{\mathbf{u}}\,.\,\hat{\mathbf{u}}) = \sum (a_1 b_2 - a_2 b_1)^2.$$

Further, putting from (4.10), (4.12) and $\hat{\mathbf{u}}\,.\,\hat{\mathbf{u}} = 1$, we get the result. //

Corollary 4.3. The vectors, given by (3.3), are *parallel* if

$$\mathbf{a} \times \mathbf{b} = \mathbf{0} \iff a_1 / b_1 = a_2 / b_2 = a_3 / b_3. \tag{4.21}$$

Proof. It follows from (4.16) that two non–null vectors **a** and **b** are parallel if the angle between them is zero. Hence, the first condition follows from (4.16) itself. Further, it follows from (4.20) that $\theta = 0$ if the numerator in the RHS of (4.20) vanishes identically, which is so if each individual term vanishes therein giving rise to the second condition. //

Theorem 4.7. For any scalar x and two vectors **a** and **b**, there hold

$$x\,(\mathbf{a} \times \mathbf{b}) = (x\,\mathbf{a}) \times \mathbf{b} = (\mathbf{a} \times x\,\mathbf{b}). \tag{4.22}$$

Proof being simple is left to the reader.

Theorem 4.8. For any two vectors **a** and **b** there hold

$$\mathbf{a}\,.\,(\mathbf{a} \times \mathbf{b}) = 0 = \mathbf{b}\,.\,(\mathbf{a} \times \mathbf{b}). \tag{4.23}$$

Proof. By definition, $\mathbf{a} \times \mathbf{b}$ is orthogonal to both **a** and **b**. Hence, the results follow from Theorem 4.1. //

Theorem 4.9. The operation of cross product is distributive over vector addition:

$$\mathbf{a} \times (\mathbf{b} + \mathbf{c}) = \mathbf{a} \times \mathbf{b} + \mathbf{a} \times \mathbf{c}. \tag{4.24}$$

Proof. Computing the cross product of $\mathbf{a} = (a_1, a_2, a_3)$ and

$$\mathbf{b} + \mathbf{c} = (b_1 + c_1, b_2 + c_2, b_3 + c_3), \tag{4.25}$$

by Eq. (4.19), we evaluate the LHS of (4.24):

$$\text{LHS} = \{a_2 (b_3 + c_3) - a_3 (b_2 + c_2)\}\hat{\mathbf{i}} + \{a_3 (b_1 + c_1) - a_1 (b_3 + c_3)\}\,\hat{\mathbf{j}}$$
$$+ \{a_1 (b_2 + c_2) - a_2 (b_1 + c_1)\}\,\hat{\mathbf{k}}\,.$$

Similarly, computing the cross product $\mathbf{a} \times \mathbf{c}$ and adding it to that given by Eq. (4.19), the sum may be seen same as above.

Example 4.1. Let $\mathbf{a}$ and $\mathbf{b}$ be two vectors with magnitudes a and b respectively. Show that

$$(\mathbf{a} + \mathbf{b}).(\mathbf{a} - \mathbf{b}) = a^2 - b^2. \tag{4.26}$$

Solution. Applying distributive law (4.15) and the commutative property of the *dot product* we prove the result. //

Example 4.2. Let $\mathbf{a}$, $\mathbf{b}$, $\mathbf{c}$ be three mutually perpendicular vectors of same magnitude. Show that their sum vector $\mathbf{a} + \mathbf{b} + \mathbf{c}$ is equally inclined to each of the vectors $\mathbf{a}$, $\mathbf{b}$, $\mathbf{c}$.

Solution. The magnitude of the vector $\mathbf{a} + \mathbf{b} + \mathbf{c}$, by Eq. (4.3), is

$$|\,\mathbf{a} + \mathbf{b} + \mathbf{c}\,| = \sqrt{\{(\mathbf{a} + \mathbf{b} + \mathbf{c}) \,.\, (\mathbf{a} + \mathbf{b} + \mathbf{c})\}} = \sqrt{(a^2 + b^2 + c^2)} = a\sqrt{3},$$

where the relations analogous to those given by Eqs. (4.4) satisfied by mutually orthogonal vectors $\mathbf{a}$, $\mathbf{b}$, $\mathbf{c}$ are used. Therefore, we get

$$(\mathbf{a} + \mathbf{b} + \mathbf{c}) \,.\, \mathbf{a} = a^2 = (a^2\sqrt{3}) \cos\theta_1 \quad \Rightarrow \quad \cos\theta_1 = 1/\sqrt{3}.$$

Similarly, we can show that $\cos\theta_2 = \cos\theta_3 = 1/\sqrt{3}$. //

Example 4.3. For any three vectors $\mathbf{a}$, $\mathbf{b}$, $\mathbf{c}$ prove that

$$\mathbf{a} \times (\mathbf{b} + \mathbf{c}) + \mathbf{b} \times (\mathbf{c} + \mathbf{a}) + \mathbf{c} \times (\mathbf{a} + \mathbf{b}) = \mathbf{0}. \tag{4.27}$$

Solution. Applying (4.24), the LHS of above identity becomes

$$\mathbf{a} \times \mathbf{b} + \mathbf{a} \times \mathbf{c} + \mathbf{b} \times \mathbf{c} + \mathbf{b} \times \mathbf{a} + \mathbf{c} \times \mathbf{a} + \mathbf{c} \times \mathbf{b},$$

which, for skew–symmetric property of cross products, vanishes. //

Example 4.4. For any vectors **a** and **b** of magnitudes a and b, prove

$$(\mathbf{a} \times \mathbf{b})^2 + (\mathbf{a} \,.\, \mathbf{b})^2 = (a\,b)^2. \qquad (4.28)$$

Solution. Squaring the results in Eqs. (4.1) and (4.16) and adding them, the LHS of Eq. (4.28) becomes

$$|\,\mathbf{a}\,|^2\,|\,\mathbf{b}\,|^2\,(\sin^2\theta + \cos^2\theta) = a^2\,b^2. \text{ //}$$

Example 4.5. For any two vectors **a** and **b**, show that

$$(\mathbf{a} - \mathbf{b}) \times (\mathbf{a} + \mathbf{b}) = 2\,(\mathbf{a} \times \mathbf{b}). \qquad (4.29)$$

Solution. Applying the distributive law (4.24), the LHS of above identity simplifies to

$$\mathbf{a} \times (\mathbf{a} + \mathbf{b}) - \mathbf{b} \times (\mathbf{a} + \mathbf{b}) = \mathbf{a} \times \mathbf{b} - \mathbf{b} \times \mathbf{a} = 2\,(\mathbf{a} \times \mathbf{b}). \text{ //}$$

§ 5. Products of Three Vectors

As seen in the previous section, the cross product of two vectors **a** and **b** is a vector. So, its further products with a third vector, say

$$\mathbf{c} = (c_1, c_2, c_3) \qquad (5.1)$$

can also be defined in the following two ways:

(i) dot product: $(\mathbf{a} \times \mathbf{b}).\mathbf{c}$, denoted by $[\mathbf{a}\ \ \mathbf{b}\ \ \mathbf{c}]$. Being a scalar, it is also called a *scalar* (*triple*) *product* of vectors **a**, **b**, **c**.

(ii) cross product: $(\mathbf{a} \times \mathbf{b}) \times \mathbf{c} \equiv \mathbf{d}$ (say) (5.2)

defined in analogy with (4.16). Being a vector, it is also called a *vector* (*triple*) *product* of vectors **a**, **b**, **c**.

(iii) As seen in § 4, the dot product of any two of the vectors **a**, **b**, **c**, say **a . b** is scalar. Hence, its product with the third vector **c** is just a scalar multiplication $(\mathbf{a} \,.\, \mathbf{b})\,\mathbf{c}$, which is already discussed in § 2 above.

5.1. Properties of [a b c]

Theorem 5.1. The dot product of three vectors **a**, **b** and **c** given by Eqs. (3.3) and (5.1) is

$$[\mathbf{a}\ \mathbf{b}\ \mathbf{c}] = \begin{vmatrix} a_1 & a_2 & a_3 \\ b_1 & b_2 & b_3 \\ c_1 & c_2 & c_3 \end{vmatrix}. \tag{5.3}$$

Proof. Forming dot product of the vectors given by Eqs. (4.19) and (5.1), we get

$$[\mathbf{a}\ \mathbf{b}\ \mathbf{c}] = (a_2 b_3 - a_3 b_2)\, c_1 + (a_3 b_1 - a_1 b_3)\, c_2 + (a_1 b_2 - a_2 b_1)\, c_3,$$

which is just the expansion of the determinant in (5.3). //

Corollary 5.1. The scalar triple product of orthonormal vectors $\hat{\mathbf{i}}$, $\hat{\mathbf{j}}$, $\hat{\mathbf{k}}$ has value 1:

$$[\hat{\mathbf{i}}\ \hat{\mathbf{j}}\ \hat{\mathbf{k}}] = 1. \tag{5.4}$$

Proof. $[\hat{\mathbf{i}}\ \hat{\mathbf{j}}\ \hat{\mathbf{k}}] = (\hat{\mathbf{i}} \times \hat{\mathbf{j}}) \,.\, \hat{\mathbf{k}} = \hat{\mathbf{k}} \,.\, \hat{\mathbf{k}} = 1$, by Eqs. (4.4) and (4.18). //

Theorem 5.2. The scalar triple product [**a** **b** **c**] gives the volume V of a parallelopiped having **a**, **b**, **c** as the coterminus edges; and it satisfies

$$V = [\mathbf{a}\ \mathbf{b}\ \mathbf{c}] = [\mathbf{b}\ \mathbf{c}\ \mathbf{a}] = [\mathbf{c}\ \mathbf{a}\ \mathbf{b}]. \tag{5.5}$$

Corollary 5.2. Three vectors are coplanar (i.e. lying in the same plane) iff their scalar triple product is zero:

$$[\mathbf{a}\ \mathbf{b}\ \mathbf{c}] = 0. \tag{5.6}$$

Proof. The result follows immediately from (5.5) as the parallelopiped formed by three coplanar vectors is of zero volume.

Alternately, if **c** lies in the plane containing **a** and **b** so it is orthogonal to **a** × **b** as the latter vector is orthogonal to that plane. This makes $(\mathbf{a} \times \mathbf{b}) \,.\, \mathbf{c} = 0$. //

Corollary 5.3. The positions of *dot* and *cross* operators are interchangeable in the product $(\mathbf{a} \times \mathbf{b}) \,.\, \mathbf{c}$.

Proof. The proof follows from the commutative property of the dot product of two vectors and the Eqs. (5.5). //

Corollary 5.4. If any two vectors in $[\mathbf{a}\ \ \mathbf{b}\ \ \mathbf{c}]$ are equal (or collinear) the scalar triple product vanishes:

$$[\mathbf{a}\ \ \mathbf{b}\ \ \mathbf{b}] = [\mathbf{a}\ \ \mathbf{a}\ \ \mathbf{c}] = [\mathbf{a}\ \ \mathbf{b}\ \ \mathbf{a}] = 0. \tag{5.7}$$

Proof. $\mathbf{a} \times \mathbf{b}$ being perpendicular to $\mathbf{b}$ (as well as to $\mathbf{a}$) its dot products with both $\mathbf{b}$ and $\mathbf{a}$ vanish:

$$(\mathbf{a} \times \mathbf{b}) \,.\, \mathbf{b} = [\mathbf{a}\ \ \mathbf{b}\ \ \mathbf{b}] = 0, \qquad (\mathbf{a} \times \mathbf{b}) \,.\, \mathbf{a} = [\mathbf{a}\ \ \mathbf{b}\ \ \mathbf{a}] = 0.$$

Similarly, $\mathbf{a} \times \mathbf{c}$ being perpendicular to $\mathbf{a}$ its dot product with $\mathbf{a}$ vanishes:

$$(\mathbf{a} \times \mathbf{c}) \,.\, \mathbf{a} = \mathbf{a} \,.\, (\mathbf{a} \times \mathbf{c}) = [\mathbf{a}\ \ \mathbf{a}\ \ \mathbf{c}] = 0. \ //$$

5.2. Properties of $(\mathbf{a} \times \mathbf{b}) \times \mathbf{c}$

As seen in § 4, the vector $\mathbf{a} \times \mathbf{b}$ is orthogonal to both $\mathbf{a}$ and $\mathbf{b}$. So, it is normal to the plane containing $\mathbf{a}$ and $\mathbf{b}$. Analogously, the vector $\mathbf{d}$ given by Eq. (5.2) is orthogonal to both $\mathbf{a} \times \mathbf{b}$ as well $\mathbf{c}$. Hence, $\mathbf{d}$ lies in the plane containing $\mathbf{a}$ and $\mathbf{b}$. Therefore, it can be linearly expressed in terms of $\mathbf{a}$ and $\mathbf{b}$:

$$\mathbf{d} = l\mathbf{a} + m\mathbf{b}, \tag{5.8}$$

where l, m are some scalars to be determined. Forming the dot product of this vector with $\mathbf{c}$ and noting that $\mathbf{d} \,.\, \mathbf{c} = 0$, we derive

$$l\,(\mathbf{a} \,.\, \mathbf{c}) + m\,(\mathbf{b} \,.\, \mathbf{c}) = 0 \ \Rightarrow\ l\,/\,(\mathbf{b} \,.\, \mathbf{c}) = -m\,/\,(\mathbf{a} \,.\, \mathbf{c}) = n \text{ (say)}$$

$$\Rightarrow \qquad l = n\,(\mathbf{b} \,.\, \mathbf{c}) \quad \text{and} \qquad m = -n\,(\mathbf{a} \,.\, \mathbf{c}).$$

Accordingly, Eq. (5.8) reduces to

$$(\mathbf{a} \times \mathbf{b}) \times \mathbf{c} = n\,(\mathbf{b} \,.\, \mathbf{c})\,\mathbf{a} - n\,(\mathbf{a} \,.\, \mathbf{c})\,\mathbf{b}. \tag{5.9}$$

Particularly, choosing $\mathbf{a} = \hat{\mathbf{i}}$ and $\mathbf{b} = \mathbf{c} = \hat{\mathbf{j}}$, where $\hat{\mathbf{i}}$, $\hat{\mathbf{j}}$ are mutually orthogonal unit vectors, Eq. (5.9) reduces to

$$(\hat{\imath} \times \hat{\jmath}) \times \hat{\jmath} = \hat{\mathbf{k}} \times \hat{\jmath} = -\hat{\imath} = n\,(\hat{\jmath}\,.\,\hat{\jmath})\,\hat{\imath} - n\,(\hat{\imath}\,.\,\hat{\jmath})\,\hat{\jmath} = n\,\hat{\imath} \Rightarrow n = -1,$$

where Eqs. (4.4) and (4.18) are used. Hence, Eq. (5.9) finally assumes the form

$$(\mathbf{a} \times \mathbf{b}) \times \mathbf{c} = (\mathbf{a}\,.\,\mathbf{c})\,\mathbf{b} - (\mathbf{b}\,.\,\mathbf{c})\,\mathbf{a}. \tag{5.10}$$

V. Differentiation of Vectors

§ 1. Derivation of a Vector

The position vector $\mathbf{r}$ of a point P is given by Eq. (3.1) of Unit IV:

$$\mathbf{r} = x\,\hat{\imath} + y\,\hat{\jmath} + z\,\hat{\mathbf{k}} \equiv (x, y, z), \tag{1.1}$$

where x, y, z are independent variables. The unit vectors $\hat{\imath}$, $\hat{\jmath}$, $\hat{\mathbf{k}}$ being fixed (along three mutually orthogonal coordinate axes), the partial derivatives of $\mathbf{r}$ with respect to (w.r.t.) x, y, z are

$$\partial\,\mathbf{r} / \partial\,x = \hat{\imath}, \qquad \partial\,\mathbf{r} / \partial\,y = \hat{\jmath}, \qquad \partial\,\mathbf{r} / \partial\,z = \hat{\mathbf{k}}. \tag{1.2}$$

The vector differential operator (also called '*del*' operator or '*nabla*')

$$\nabla \equiv \hat{\imath}\,(\partial / \partial\,x) + \hat{\jmath}\,(\partial / \partial\,y) + \hat{\mathbf{k}}\,(\partial / \partial\,z) \tag{1.3}$$

gives rise to *three* different kinds of derivatives of a scalar f and vector function $\mathbf{u}$:

$$\nabla f = \{\,\hat{\imath}\,(\partial/\partial x) + \hat{\jmath}\,(\partial/\partial y) + \hat{\mathbf{k}}\,(\partial/\partial z)\}\,f$$

$$= \hat{\imath}\,(\partial f/\partial x) + \hat{\jmath}\,(\partial f/\partial y) + \hat{\mathbf{k}}\,(\partial f/\partial z), \tag{1.4}$$

$$\nabla\,.\,\mathbf{u} = \{\hat{\imath}\,(\partial/\partial x) + \hat{\jmath}\,(\partial/\partial y) + \hat{\mathbf{k}}\,(\partial/\partial z)\}\,.\,\mathbf{u}$$

$$= \hat{\imath}\,.\,(\partial\mathbf{u}/\partial x) + \hat{\jmath}\,.\,(\partial\mathbf{u}/\partial y) + \hat{\mathbf{k}}\,.\,(\partial\mathbf{u}/\partial z), \tag{1.5}$$

and

$$\nabla \times \mathbf{u} = \{\,\hat{\imath}\,(\partial/\partial x) + \hat{\jmath}\,(\partial/\partial y) + \hat{\mathbf{k}}\,(\partial/\partial z)\} \times \mathbf{u}$$

$$= \hat{\imath} \times (\partial\mathbf{u} / \partial x) + \hat{\jmath} \times (\partial\mathbf{u} / \partial y) + \hat{\mathbf{k}} \times (\partial\mathbf{u} / \partial z). \tag{1.6}$$

The first of these being a vector is called the *gradient* of (scalar) function f, the second one is a scalar function and is called the *divergence* of (the vector) **u** and the third one is again a vector called the *curl* (or *rot*) of the vector **u**. These derivatives are briefly denoted as grad f, div **u** and curl **u** respectively.

Theorem 1.1. For the vector field $\mathbf{u} = u_1\,\hat{\mathbf{i}} + u_2\,\hat{\mathbf{j}} + u_3\,\hat{\mathbf{k}}$, we have

$$\nabla \,.\, \mathbf{u} \;=\; \partial u_1/\partial x + \partial u_2/\partial y + \partial u_3/\partial z, \tag{1.7a}$$

and

$$\nabla \times \mathbf{u} \;=\; \begin{vmatrix} \hat{\mathbf{i}} & \hat{\mathbf{j}} & \hat{\mathbf{k}} \\ \partial/\partial x & \partial/\partial y & \partial/\partial z \\ u_1 & u_2 & u_3 \end{vmatrix}$$

$$= (\partial u_3/\partial y - \partial u_2/\partial z)\,\hat{\mathbf{i}} + (\partial u_1/\partial z - \partial u_3/\partial x)\,\hat{\mathbf{j}} + (\partial u_2/\partial x - \partial u_1/\partial y)\,\hat{\mathbf{k}}\,. \tag{1.7b}$$

Proof. Computing the dot and cross product of two vectors ∇ and **u**, and application of the Eq. (4.19) of Unit IV yields the result.

Example 1.1. We have

$$\text{div}\,\mathbf{r} \;=\; 3, \qquad \text{and} \qquad \text{curl}\,\mathbf{r} \;=\; \mathbf{0}. \tag{1.8}$$

Solution. Putting from Eq. (1.2) in Eq. (1.7), and using Eqs. (4.4) of Unit IV, we easily derive the results. //

§ 2. Some Identities

$$\nabla\,(u\,v) = (\nabla u)\,v + u\,(\nabla v);\ \text{i.e. grad}\,(u\,v) = (\text{grad}\,u)\,v + u\,(\text{grad}\,v), \tag{2.1}$$

$$\nabla\,(\mathbf{a}\,.\,\mathbf{b}) \;= \mathbf{a} \times (\nabla \times \mathbf{b}) + \mathbf{b} \times (\nabla \times \mathbf{a}) + (\mathbf{a}\,.\,\nabla)\,\mathbf{b} + (\mathbf{b}\,.\,\nabla)\,\mathbf{a},$$

i.e.

$$\text{grad}\,(\mathbf{a}\,.\,\mathbf{b}) = \mathbf{a} \times (\text{curl}\,\mathbf{b}) + \mathbf{b} \times (\text{curl}\,\mathbf{a}) + (\mathbf{a}\,.\,\nabla)\,\mathbf{b} + (\mathbf{b}.\,\nabla)\,\mathbf{a}, \tag{2.2}$$

$$\nabla.\,(u\,\mathbf{a}) \;= (\nabla u).\,\mathbf{a} + u\,(\nabla.\,\mathbf{a});\ \text{i.e. div}\,(u\,\mathbf{a}) = (\text{grad}\,u).\,\mathbf{a} + u\,(\text{div}\,\mathbf{a}), \tag{2.3}$$

$$\nabla\times\,(u\,\mathbf{a}) = (\nabla u) \times \mathbf{a} + u\,(\nabla\times\,\mathbf{a});\ \text{i.e. curl}\,(u\,\mathbf{a}) = (\text{grad}\,u) \times \mathbf{a} + u\,(\text{curl}\,\mathbf{a}), \tag{2.4}$$

$$\nabla.\,(\mathbf{a} \times \mathbf{b}) \;=\; (\nabla \times \mathbf{a})\,.\,\mathbf{b} - \mathbf{a}\,.\,(\nabla \times \mathbf{b}),$$

i.e.

$$\text{div}\,(\mathbf{a} \times \mathbf{b}) = (\text{curl}\,\mathbf{a}) \,.\, \mathbf{b} - \mathbf{a} \,.\, (\text{curl}\,\mathbf{b}), \tag{2.5}$$

$$\nabla \times (\mathbf{a} \times \mathbf{b}) = \mathbf{a}\,(\nabla .\, \mathbf{b}) - \mathbf{b}\,(\nabla .\, \mathbf{a}) + (\mathbf{b} \,.\, \nabla)\,\mathbf{a} - (\mathbf{a} \,.\, \nabla)\,\mathbf{b},$$

i.e.

$$\text{curl}\,(\mathbf{a} \times \mathbf{b}) = \mathbf{a}\,(\text{div}\,\mathbf{b}) - \mathbf{b}\,(\text{div}\,\mathbf{a}) + (\mathbf{b} \,.\, \nabla)\,\mathbf{a} - (\mathbf{a} \,.\, \nabla)\,\mathbf{b}. \tag{2.6}$$

CHAPTER 2

ORDINARY DIFFERENTIAL EQUATIONS

§ 1. Introduction

The present chapter is a pre-requisite for the subsequent topics discussed in the book. So, a brief (*revision*) course of ordinary differential equations is given here. Equations involving one (or more) dependent variable(s), independent variable(s) and derivatives of dependent variable(s) with respect to the independent variable(s) are called *differential equations*. Presently, we consider ordinary differential equations involving *one* independent variable, say x, *one* dependent variable, say y (which is a function of x), and derivatives of y with respect to x. When the differential equation is written in a rational form, the highest order of differentiation of y w.r.t. x appearing in the equation defines the *order* of the differential equation; while the highest degree of dy/dx defines the *degree* of the differential equation. Thus, the differential equation

$$dy / dx = f(x, y) \tag{1.1}$$

is an *ordinary differential equation* of *first degree* and of *first order* in two variables: x (independent) and y (dependent). Also, a differential equation is said to be *linear* when both the dependent variable y and its derivative dy/dx are of the first degree. Thus,

$$P(x). (dy / dx) + Q(x). y = R(x), \tag{1.2}$$

where P, Q, R are some functions of x alone, is a linear differential equation. Any relation between the dependent variable and the independent variable satisfying the differential equation is called a *solution* (or *integral*) of the differential equation. A solution having the same number of arbitrary constants equal to the order of the differential equation is called a *general solution* (or *complete integral* or *complete primitive*). In the next section, we deal with the differential equations of the type (1.1) and discuss different methods for their integration.

On contrary, there are differential equations of the *first* order but of degree more than one. For example, if $F(x, y, p)$ is a function of any degree in its variables: (independent) x, (dependent) y and $p \equiv dy/dx$, then a relation

$$F(x, y, p) = 0 \tag{1.3}$$

expresses a differential equation of above type. The solution of such differential equations will be discussed in the Section 3.

The linear differential equations of *any order* (but of *degree one*) with constant coefficients will be discussed in the Section 4. The following equation

$$a_0 (d^n y / dx^n) + a_1 (d^{n-1} y / dx^{n-1}) + a_2 (d^{n-2} y / dx^{n-2}) + \dots$$

$$+ a_{n-1} (dy / dx) + a_n y = R(x) \qquad (1.4)$$

represents a *linear* differential equation of *first degree* but of nth *order.*

§ 2. Differential Equations of First Order and First Degree

As seen in the preceding Section, the equation (1.1) represents a differential equation of first order and first degree. Every such differential equation need not be solvable. Methods of solution of some of the special forms of this equation are suggested below.

2.1. Separable variables form: Let the equation (1.1) be of the form

$$dy / dx = f_1(x) / f_2(y), \qquad \text{or} \qquad f_1(x).\, dx - f_2(y).\, dy = 0, \quad (2.1)$$

where the coefficients $f_1(x)$ and $f_2(y)$ are functions of the respective variables alone. Integrating each term in Eq. (2.1) separately w.r.t. their respective variables, we get

$$\int f_1(x).\, dx - \int f_2(y).\, dy = c, \qquad (2.2)$$

c being an arbitrary constant of integration.

2.2. Homogeneous form: A differential equation of the form

$$dy / dx = f_1(x, y) / f_2(x, y), \qquad (2.3)$$

where f_1, f_2 are some homogeneous functions of both variables x and y and have the same degree is said to be in a *homogeneous form.* Setting $y = v.\, x$ so that

$$dy / dx = v + x.\, dv / dx,$$

where v is a new variable depending on x, reduces the Eq. (2.3) to

$$v + x.\, dv/dx = f_1\,(x, v.x)\,/\,f_2\,(x, v.x) \equiv h\,(v) \;\Rightarrow\; dv\,/\{h\,(v) - v\} = dx/x.$$

This is a differential equation in v and x, where the variables are separated. Integrating it term wise w.r.t. the respective variables, we get

$$\int [1\,/\,\{h\,(v) - v\}.\, dv] \;=\; \ln x + c.$$

Finally, v is replaced by $y\,/\,x$ in above equation to yield the solution of Eq. (2.3).

2.3. Reducible to homogeneous form: A differential equation of the form

$$dy\,/\,dx \;=\; (a_1\,x + b_1\,y + c_1)\,/\,(a_2\,x + b_2\,y + c_2) \qquad (2.4)$$

can be reduced to a homogeneous form by introducing new set of variables ξ , η and some constants h and k :

$$x = \xi + h,\;\; y = \eta + k \;\;\Rightarrow\;\; dy\,/\,dx \;=\; \{d\,(\eta + k)\,/\,d\xi\}\,(d\xi/dx) \;=\; d\eta\,/\,d\xi.$$

Thus, the differential equation (2.4) reduces to

$$d\eta/d\xi = \{a_1\,\xi + b_1\eta + (a_1h + b_1k + c_1)\}/\{a_2\,\xi + b_2\eta + (a_2h + b_2k + c_2)\}.$$

Choosing the constants h and k so as to satisfy the simultaneous equations

$$a_1\,h + b_1\,k + c_1 \;=\; 0, \qquad \text{and} \qquad a_2\,h + b_2\,k + c_2 \;=\; 0,$$

the differential equation further simplifies to

$$d\eta\,/\,d\xi \;=\; (a_1\,\xi + b_1\,\eta)\,/\,(a_2\,\xi + b_2\,\eta). \qquad (2.5)$$

This is in a homogeneous form given by Eq. (2.3): both $a_1\,\xi + b_1\,\eta$ and $a_2\,\xi + b_2\,\eta$ being homogeneous functions of ξ and η of degree *one*. Setting $\eta = v.\,\xi$ and proceeding as in the preceding Sub-section we may find the solution of Eq. (2.5).

2.4. A linear form: The equation (1.2) represents a most general linear differential equation in variables x and y. Dividing it by $P\,(x)$ we may have its special form:

$$dy / dx + P_1 (x). y = R_1 (x), \tag{2.6}$$

where $P_1 \equiv Q/P$ and $R_1 \equiv R/P$. It can be integrated by multiplying by a factor $e^{\int P1\, dx}$ (called the *integrating factor* of the differential equation):

$$e^{\int P1\, dx} (dy / dx) + (P_1. e^{\int P1\, dx}) y = R_1. e^{\int P1\, dx}.$$

The LHS of above equation is the derivative of the product function $y. e^{\int P1\, dx}$. Hence, the equation is directly integrable and its solution is:

$$(e^{\int P1\, dx}) y = \int (R_1. e^{\int P1\, dx})\, dx + c$$

$\Rightarrow$

$$y = (e^{-\int P1\, dx}). \{ \int (R_1. e^{\int P1\, dx})\, dx + c \}. \tag{2.7}$$

2.5. Reducible to a linear form: *Bernoulli*'s equation

$$dy / dx + P_1 (x). y = R_1 (x). y^n \tag{2.8}$$

can be reduced to a linear form as follows. Dividing it by y^n:

$$y^{-n}. (dy / dx) + P_1 (x). y^{1-n} = R_1 (x),$$

and putting $y^{1-n} = v$ so that $(1 - n) y^{-n} (dy / dx) = dv / dx$ above equation reduces to

$$dv / dx + (1 - n). P_1. v = (1 - n) R_1,$$

which is of the form as in Eq. (2.6).

Also, a differential equation of the form given by

$$f'(y). (dy / dx) + P_1 (x). f(y) = R_1 (x), \tag{2.9}$$

where $f'(y) \equiv df / dy$, can be reduced to a linear form. Putting

$$f(y) = v \quad \text{so that} \quad f'(y). (dy / dx) = dv / dx,$$

above equation reduces to

$$dv / dx + P_1. v = R_1,$$

which is of the form as in Eq. (2.6).

2.6. Exact form: A differential equation obtainable by differentiating its solution directly (without performing any other mathematical operations) is called *exact*. A necessary and sufficient condition for the differential equation (2.3) modified as:

$$f_1\,(x, y).\,dx + f_2\,(x, y).\,dy = 0, \tag{2.10}$$

to be exact is

$$\partial f_1 / \partial y = \partial f_2 / \partial x. \tag{2.11}$$

The following rules to have the integrating factors of a differential equation are noteworthy:

2.6.1. When the equation (2.10) is homogeneous in its variables x and y, and

$$F\,(x, y) \equiv f_1.\,x + f_2.\,y \neq 0,$$

then the integrating factor is

$$1 / F\,(x, y) \equiv 1 / (f_1.\,x + f_2.\,y)\,. \tag{2.12}$$

2.6.2. When the equation (2.10) is of the form

$$f_1\,(x.\,y).\,y.\,dx + f_2\,(x.\,y).\,x.\,dy = 0, \tag{2.13}$$

and

$$F\,(x, y) \equiv (f_1.\,y).\,x - (f_2.\,x).\,y = (f_1 - f_2).\,x\,y \neq 0,$$

then the integrating factor is

$$1 / F(x, y) \equiv 1 / (f_1 - f_2).\,x\,y. \tag{2.14}$$

2.6.3. When

$$(\partial f_1 / \partial y - \partial f_2 / \partial x) / f_2 \equiv F\,(x) \tag{2.15}$$

is a function of x alone the integrating factor is $e^{\int F\,(x)\,dx}$.

2.6.4. When

$$(\partial f_1 / \partial y - \partial f_2 / \partial x) / f_1 \equiv -\,G\,(y) \tag{2.16}$$

is a function of y alone the integrating factor is $e^{\int G\,(y)\,dy}$.

2.6.5. When the equation (2.10) is of the form

$$x^a y^b.(a_1.y.dx + b_1.x.dy) + x^h y^k.(h_1.y.dx + k_1.x.dy) = 0, \quad (2.17)$$

where $a, b, a_1, b_1, h, k, h_1, k_1$ are all constants, then the integrating factor is $x^l y^m$. The values of l, m are found by application of the condition given by Eq. (2.11) for exactness.

2.7. Change of variables: When the variables (both dependent and independent) are transformed, some of the differential equations reduce to any of the forms discussed above. As such, they become integrable.

Example 2.1. Obtain an integral of the differential equation

$$(a^2 - 2xy - y^2)\,dx - (x + y)^2\,dy = 0. \quad (2.18)$$

Solution. Comparing it with Eq. (2.10), we note that the condition given by Eq. (2.11) is satisfied:

$$\partial f_1 / \partial y = -2\,(x + y) = \partial f_2 / \partial x,$$

making the equation in exact form. Integrating f_1 and f_2 partially w.r.t. x and y respectively:

$$\int_x f_1\,dx = \int_x (a^2 - 2x\,y - y^2)\,dx = a^2 x - x^2 y - x y^2,$$

and

$$\int_y f_2\,dy = -\int_y (x^2 + 2x\,y + y^2)\,dy = -(x^2 y + x y^2 + y^3/3).$$

Retaining all the terms in the integral of f_1 and only those in the integral of f_2 not repeating earlier a solution of Eq. (2.18) is found as

$$a^2 x - x^2 y - x y^2 - y^3/3 = b,$$

b being a constant of integration. //

Example 2.2. Finding a suitable integrating factor solve the equation

$$(x^2 y - 2xy^2)\,dx - (x^3 - 3x^2 y)\,dy = 0. \quad (2.19)$$

Solution. Comparing it with Eq. (2.10), we note that

$$\partial f_1 / \partial y = x^2 - 4\,x\,y \quad \text{and} \quad \partial f_2 / \partial x = -3x^2 + 6x\,y$$

are unequal implying the non-exact nature of the equation. However, it is comparable with Case 2.6.1. For $f_1\, x + f_2\, y = x^2 y^2 \neq 0$ and following Eq. (2.12), the integrating factor of the equation becomes $1/x^2 y^2$. Multiplication by the same makes the equation in exact form:

$$(1/y - 2/x)\,dx + (-x/y^2 + 3/y)\,dy = 0. \qquad (2.20)$$

Thus, integrating the coefficients of dx and dy in above (exact) differential equation partially w.r.t. x and y:

$$\int_x f_1\,dx = \int_x (1/y - 2/x)\,dx = x/y - 2.\ln x,$$

and

$$\int_y f_2\,dy = \int_y (-x/y^2 + 3/y)\,dy = x/y + 3.\ln y.$$

Hence, a solution of Eq. (2.20) is

$$x/y - 2.\ln x + 3.\ln y = \ln a; \qquad \text{or,} \qquad y^3.\,e^{x/y} = a\,x^2.\ //$$

Example 2.3. Finding a suitable integrating factor solve the differential equation

$$(1 + x\,y)\,y.\,dx + (1 - x\,y)\,x.\,dy = 0. \qquad (2.21)$$

Solution. Comparing it with Eq. (2.10), we note that

$$\partial f_1/\partial y = 1 + 2x\,y \qquad \text{and} \qquad \partial f_2/\partial x = 1 - 2xy$$

are unequal implying the non-exact nature of the equation. However, it is of the form of equation (2.13). Thus, for

$$\{f_1(x\,y) - f_2\,(x\,y)\}\;x\,y = 2\,x^2 y^2 \neq 0,$$

and following Eq. (2.14), the integrating factor of the equation is $1/2x^2.\ y^2$. Multiplication by the same makes the equation in exact form:

$$(1/x^2 y + 1/x)\,dx + (1/x\,y^2 - 1/y)\,dy = 0. \qquad (2.22)$$

Integrating the coefficients of dx and dy in above (exact) differential equation partially w.r.t. x and y:

$$\int_x (1/x^2 y + 1/x)\, dx = -1/xy + \ln x,$$

and

$$\int_y (1/xy^2 - 1/y)\, dy = -1/xy - \ln y.$$

Hence, a solution of Eq. (2.21) is

$$-1/xy + \ln x - \ln y = a; \quad \text{or,} \quad \ln(x/y) - 1/xy = a. //$$

Example 2.4. Finding a suitable integrating factor solve equation

$$(3x^2 y^4 + 2xy)\, dx + (2x^3 y^3 - x^2)\, dy = 0. \tag{2.23}$$

Solution. Comparing it with Eq. (2.10), we note that

$$\partial f_1 / \partial y = 12x^2 y^3 + 2x \quad \text{and} \quad \partial f_2 / \partial x = 6x^2 y^3 - 2x$$

are unequal implying the non-exact nature of the equation. However,

$$(\partial f_1 / \partial y - \partial f_2 / \partial x) / f_1 = (6x^2 y^3 + 4x) / (3x^2 y^4 + 2xy) = 2/y$$

being a function of y alone, the integrating factor of Eq. (2.23), following Case 2.6.4, is

$$\exp\{-2 \int (1/y)\, dy\} = \exp(-2.\ln y) = 1/y^2.$$

Multiplication by the same makes the equation (2.23) in exact form:

$$(3x^2 y^2 + 2x/y)\, dx + (2x^3 y - x^2/y^2)\, dy = 0.$$

Integrating the coefficients of dx and dy in above equation partially w.r.t. x and y:

$$\int_x (3x^2 y^2 + 2x/y)\, dx = x^3 y^2 + x^2/y,$$

and

$$\int_y (2x^3 y - x^2/y^2)\, dy = x^3 y^2 + x^2/y.$$

Hence, a solution of Eq. (2.23) is

$$x^3 y^2 + x^2/y = a. //$$

Example 2.5. Finding a suitable integrating factor solve the differential equation

$$(2y.\, dx + 3x.\, dy) + 2x\, y\, (3y.\, dx + 4x.\, dy) = 0. \tag{2.24}$$

Solution. Collecting the coefficients of dx and dy :

$$(2y + 6\, x\, y^2)\, dx + (3x + 8\, x^2 y)\, dy = 0, \tag{2.24a}$$

and comparing the resulting equation with Eq. (2.10), we note that

$$\partial f_1 / \partial y = 2 + 12\, x\, y \quad \text{and} \quad \partial f_2 / \partial x = 3 + 16\, xy$$

are unequal implying the non-exact nature of the equation. Following Case 2.6.5, if $x^l\, y^m$ is an integrating factor of the equation (2.24a), its multiplication turns the equation in exact form:

$$x^l\, y^m\, \{(2y + 6\, x\, y^2)\, dx + (3x + 8\, x^2\, y)\, dy\} = 0,$$

or

$$(2x^l\, y^{m+1} + 6x^{l+1}\, y^{m+2})\, dx + (3x^{l+1}\, y^m + 8x^{l+2}\, y^{m+1})\, dy = 0. \tag{2.25}$$

Hence,

$$\partial\, (2x^l\, y^{m+1} + 6x^{l+1}\, y^{m+2})/\partial y = 2\, (m + 1)\, x^l\, y^m + 6\, (m + 2)\, x^{l+1}\, y^{m+1},$$

and

$$\partial\, (3x^{l+1}\, y^m + 8x^{l+2}\, y^{m+1})/\partial x = 3\, (l + 1)\, x^l\, y^m + 8\, (l + 2)\, x^{l+1}\, y^{m+1}$$

should be equal. Equating the coefficients of like terms, there result two simultaneous linear relations in l and m:

$$2\, (m + 1) = 3\, (l + 1) \qquad \text{and} \qquad 6\, (m + 2) = 8\, (l + 2).$$

Solving these relations we get $l = 1$ and $m = 2$ implying the integrating factor $x\, y^2$ of the differential equation (2.24a). As such, the exact differential equation (2.25) then simplifies to

$$(2x\, y^3 + 6x^2\, y^4)\, dx + (3x^2\, y^2 + 8x^3\, y^3)\, dy = 0 \tag{2.25a}$$

Integrating the coefficients of dx and dy in above equation partially w.r.t. x and y :

$$\int_x (2x\, y^3 + 6x^2\, y^4)\, dx = x^2\, y^3 + 2x^3\, y^4,$$

$$\int_y (3x^2 y^2 + 8x^3 y^3)\, dy = x^2 y^3 + 2x^3 y^4.$$

Hence, a solution of Eq. (2.25a) is $x^2 y^3 + 2\, x^3 y^4 = a$. //

Example 2.6. Solve the following differential equation by a suitable change of variables:

$$(x + y)^2 \, (dy \,/\, dx) = a^2. \tag{2.26}$$

Solution. Putting $x + y = t$, so that $dy/dx = dt/dx - 1$, the differential equation transforms as

$$t^2 \,(dt \,/dx - 1) = a^2 \quad \Rightarrow \quad dt \,/dx = 1 + a^2 \,/\, t^2 = (t^2 + a^2) \,/\, t^2.$$

Separating the variables:

$$dx = \{t^2 \,/\, (t^2 + a^2)\} \, dt = \{1 - a^2 \,/\, (t^2 + a^2)\} \, dt,$$

and integrating it w.r.t. respective variables, we get

$$x = t - a.\tan^{-1} (t \,/a) + b = x + y - a.\tan^{-1} \{(x + y)/a\} + b,$$

or

$$y + b = a.\tan^{-1} \{(x + y) \,/\, a\}. \; //$$

§ 3. Differential Equations of First Order and of Any Degree

As introduced in the §1, the equation (1.3) represents a first order differential equation of some degree. The following three forms are suggested for achieving solution of some of the differential equations of this type.

3.1. Equations solvable for *p*: Let the equation (1.3) be of nth degree in p and have the form

$$\{p - f_1 (x, y)\}.\{p - f_2 (x, y)\} \;\ldots\; \{p - f_n (x, y)\} = 0. \tag{3.1}$$

Thus, there result n differential equations each of the form as in Eq. (1.1):

$$p \equiv dy \,/\, dx = f_i (x, y), \; i = 1, 2, \ldots, n. \tag{3.2}$$

Solutions of these equations are discussed in the preceding Section. Let Eq. (3.2) have solutions

$$F_i\,(x, y, c_i) = 0, \; i = 1, 2, \ldots, n \tag{3.3}$$

giving rise to a general solution

$$F_1\,(x, y, c_1).\, F_2\,(x, y, c_2) \ldots F_n\,(x, y, c_n) = 0.$$

Since the original differential equation (3.1) is of first order only its complete integral should contain only one arbitrary constant. So, without loss of generality, above n constants can be replaced by a single constant, say c. Thus, above general solution reduces to

$$F_1\,(x, y, c).\, F_2\,(x, y, c) \ldots F_n\,(x, y, c) = 0. \tag{3.4}$$

3.2. Equations solvable for y: When the equation (1.3) is solvable for y, we have

$$y = f\,(x, p). \tag{3.5}$$

Its differentiation w.r.t. x yields a relation

$$p = df/\,dx = \{(\partial f/\,\partial x) + (\partial f/\partial p).(dp/dx)\} \equiv \varphi\,(x, p, dp/dx),$$

which is a differential equation in the (independent) variable x and (dependent) variable p. It can be integrated by some method discussed in the preceding Section. Let its solution be

$$F\,(x, p, c) = 0. \tag{3.6}$$

Elimination of p between Eqs. (3.5) and (3.6) yields the desired solution. In case, this elimination is not feasible, we solve Eq. (3.6) for x:

$$x = g\,(y, p, c), \tag{3.7}$$

and leave the solution in the form given by Eqs. (3.5) and (3.7) treating p as a parameter. This method is especially useful when the differential equation does not contain the independent variable x.

3.3. Equations solvable for x : Let the equation (1.3) be solvable for x:

$$x = f\,(y, p). \tag{3.8}$$

Differentiating it w.r.t. y, we get

$$1/p \equiv dx/dy = df/dy = \{(\partial f/\partial y) + (\partial f/\partial p).(dp/dy)\} \equiv \varphi\,(y, p, dp/dy),$$

which is a differential equation in the (independent) variable y and (dependent) variable p. It can be integrated by some method discussed in the preceding Section. Let its solution be

$$F\,(y, p, c) = 0. \tag{3.9}$$

Elimination of p between Eqs. (3.8) and (3.9) yields the desired solution. In case, this elimination is not feasible, we solve Eq. (3.9) for y:

$$y = g\,(x, p, c), \tag{3.10}$$

and leave the solution in the form given by Eqs. (3.8) and (3.10), where p acts as a parameter.

3.4. Clairaut's form: The equation of the form

$$y = p.\,x + f(p) \tag{3.11}$$

is known after F.C. Clairaut. Its differentiation w.r.t. x yields

$$p = p + \{x + (df/dp)\}.(dp/dx)$$

$\Rightarrow \quad dp/dx = 0 \quad$ (3.12a); $\qquad$ or, $\qquad df/dp + x = 0. \quad$ (3.12b)

The first alternative has integral $p = c$; which, in conjunction with Eq. (3.11), yields the desired general solution of Eq. (3.11):

$$y = c.\,x + f(c). \tag{3.13}$$

On the other hand, Eq. (3.12b) represents a differential equation in the variables x (independent) and p (dependent), where the variables are separated. So, its integral is

$$f(p) + x^2/2 = 0. \tag{3.14}$$

Elimination of p between Eqs. (3.11) and (3.14) yields a solution which does not contain any arbitrary constant and is neither a particular case of the solution given by Eq. (3.13). Such a solution is termed as a *singular solution* of the differential equation (3.11).

3.4.1. A more general form: A more general equation of the form

$$y = x.f(p) + F(p) \tag{3.15}$$

can also be solved similarly. Its differentiation w.r.t. x yields

$$p = f(p) + \{x.(df/dp) + dF/dp\}.(dp/dx),$$

or

$$\{p - f(p)\}.(dx/dp) - x.(df/dp) = dF/dp. \tag{3.16}$$

A comparison of this equation with Eq. (1.2) shows that it is linear in the variables p (treated as an independent variable) and the dependent variable x. The solution of such equations has been discussed before.

Example 3.1. Solve the following differential equations, where $p = dy/dx$:

(i) $xy\,p^2 + p\,(3x^2 - 2y^2) - 6xy = 0$, **(ii)** $x^2p^2 - 2xyp + y^2 = x^2y^2 + x^4$,

(iii) $y + x\,p = x^4\,p^2$, **(iv)** $y = p.\sin p + \cos p$,

(v) $p^3 - p\,(y+3) + x = 0$, **(vi)** $p = \ln(p\,x - y)$.

Solution. (i) Factorizing the expression on LHS and solving the equation for p, we get

$$(px - 2y).(py + 3x) = 0 \Rightarrow p = 2y/x,\ p = -3x/y. \tag{3.17}$$

Separating the variables and integrating these differential equations,

$$\int (1/y)\,dy = \int (2/x)\,dx + \ln a, \quad \text{or,} \quad \ln y = \ln x^2 + \ln a \Rightarrow y = a.x^2,$$

and

$$\int y\,dy + 3\int x\,dx = a/2, \qquad \text{or,} \qquad y^2 + 3x^2 = a.$$

Thus, the combined general solution of the differential equation is

$$(y - a.x^2).(y^2 + 3x^2 - a) = 0.$$

It may be noted that the same arbitrary constant of integration a is taken in the solution of both the differential equations (3.17) as the given differential equation is of first order. So, its general solution has to involve only one arbitrary constant of integration.

(ii) Solving the equation for p, we get

$$p \equiv dy/dx = \{y \pm x\sqrt{(x^2+y^2)}\}/x.$$

The substitution

$$y = vx \quad \Rightarrow \quad dy/dx = v + x\,(dv/dx)$$

reduces above equation as

$$v + x\,(dv/dx) = v \pm x\sqrt{(1+v^2)} \Rightarrow \quad dv/\sqrt{(1+v^2)} = \pm\, dx.$$

Integrating term wise we get

$$\sinh^{-1} v = \pm(x+a) \Rightarrow v \equiv y/x = \sinh\{\pm(x+a)\} = \pm\sinh(x+a).$$

Thus, the combined general solution is

$$\{y - x.\sinh(x+a)\}\{y + x.\sinh(x+a)\} = y^2 - x^2.\sinh^2(x+a) = 0.$$

(iii) Solving the equation for y and differentiating it w.r.t. x, we get

$$dy/dx \equiv p = 4x^3p^2 - p + (2px^4 - x)\,(dp/dx),$$

or,

$$2p\,(2px^3 - 1) + x\,(2px^3 - 1)\,(dp/dx) = 0,$$

or, on division by the factor $2px^3 - 1$, we get the differential equation $2dx/x + dp/p = 0$. Integrating it term by term, there results a general solution

$$\ln x^2 + \ln p = \ln a; \qquad \text{or,} \qquad px^2 = a \Rightarrow p = a/x^2.$$

Eliminating p in the given differential equation, we get the general solution

$$y + a/x = a^2 \qquad \Rightarrow \qquad xy + a = a^2.x.$$

(iv) Differentiating the given equation w.r.t. x, we get

$$dy/dx \equiv p = p.\cos p.(dp/dx); \text{ or, on division by } p, \quad dx/dp = \cos p.$$

Integrating it term wise, there results a value for x in terms of p: $x = \sin p + a$. The same in association with the given equation constitute a general solution, where p acts as a parameter.

(v) Solving the equation for x and differentiating it w.r.t. y, we get

$$dx/dy \equiv 1/p = p + (y + 3 - 3p^2)\,(dp/dy)$$

or,

$$(p - 1/p)\,(dy/dp) + y = 3\,(p^2 - 1),$$

or, on multiplication by $p / (p^2 - 1)$,

$$dy / dp + \{p / (p^2 - 1)\}\, y = 3p.$$

This is a linear differential equation of first order and first degree in (dependent) variable y and (independent) variable p and can be solved by the method given in Sub-section 2.4. Its integrating factor is

$$\exp\,[\textstyle\int \{p / (p^2 - 1)\}\, dp] = \exp\,\{(1/2) \ln (p^2 - 1)\} = \sqrt{(p^2 - 1)},$$

and the solution is

$$y \sqrt{(p^2 - 1)} = \int 3p \sqrt{(p^2 - 1)}\, dp.$$

Putting $p^2 - 1 = t^2$ so that $p\, dp = t\, dt$, above solution becomes

$$y \sqrt{(p^2 - 1)} = 3 \int t^2\, dt + a = t^3 + a = (p^2 - 1)^{3/2} + a,$$

or,

$$y = (p^2 - 1) + a / \sqrt{(p^2 - 1)}.$$

Putting for y in the given differential equation, we also evaluate x:

$$x = 3p - p^3 + p\,\{p^2 - 1 + a / \sqrt{(p^2 - 1)}\} = 2p + ap / \sqrt{(p^2 - 1)}.$$

The values of x and y (expressed in terms of the parameter p) constitute a general solution of the given differential equation.

(vi) Rewriting the equation as $e^p = px - y \Rightarrow y = px - e^p$, which is in Clairaut's form having a general solution

$$y = ax - e^a, \text{ or, } \quad ax - y = e^a \quad \Rightarrow \quad \ln (ax - y) = a. \;//$$

§ 4. Linear Differential Equations of Any Order with Constant Coefficients

A linear differential equation of nth *order* and of *degree one* in the variables x (independent) and y (dependent) is represented by the equation (1.4). Let the coefficients $a_0, a_1, a_2, \ldots, a_n$ therein be constant. The solution of such differential equations is discussed in the following.

First, we consider a special form of such equation with vanishing function $R(x)$:

$$a_0 (d^n y / dx^n) + a_1 (d^{n-1} y / dx^{n-1}) + a_2 (d^{n-2} y / dx^{n-2}) + \ldots$$

$$+ a_{n-1} (dy / dx) + a_n y = 0. \tag{4.1}$$

Let $y = e^{mx}$ be a solution of this equation, then there holds

$$e^{mx} .(a_0 m^n + a_1 m^{n-1} + a_2 m^{n-2} + \ldots + a_{n-1} m + a_n) = 0,$$

implying that m is a root of the nth degree (algebraic) equation

$$a_0 m^n + a_1 m^{n-1} + a_2 m^{n-2} + \ldots + a_{n-1} m + a_n = 0. \tag{4.2}$$

This equation is called an *auxiliary equation* or *indicial equation* of the differential equation. There arise different situations for the roots of this auxiliary equation.

4.1. All real and distinct roots: Let $m_1, m_2, \ldots, m_n$ be all real and distinct roots of the auxiliary equation giving rise to n independent solutions $e^{m_1 x}, e^{m_2 x}, \ldots, e^{m_n x}$ or a complete solution (called the *complementary function*)

$$y = A_1 e^{m_1 x} + A_2 e^{m_2 x} + \ldots + A_n e^{m_n x}. \tag{4.3}$$

4.2. Some equal roots: Let the auxiliary equation have *two* equal real roots, say $m_1 = m_2$ and $n - 2$ distinct (real) roots $m_3, \ldots, m_n$. Correspondingly, the solution given by Eq. (4.3) becomes

$$y = (A_1 + A_2) e^{m_1 x} + A_3 e^{m_3 x} + \ldots + A_n e^{m_n x}.$$

Since the sum $A_1 + A_2$ of two arbitrary constants can be replaced by a single constant, say B, the number of arbitrary constants of integration reduces from n to $n-1$. As such, above solution cannot represent a general solution of the nth order differential equation (4.1). To overcome this difficulty, we proceed as follows for a general solution of such a differential equation. Beginning with a simple (*second order*) differential equation

$$a_0 (d^2 y / dx^2) + a_1 (dy / dx) + a_2 y = 0, \tag{4.4}$$

or, setting $a_1 = -2a_0 m_1$ and $a_2 = a_0 m_1^2$,

$$a_0 \{d^2 y / dx^2 - 2m_1 (dy/dx) + m_1^2 y\} \equiv a_0 (d / dx - m_1)^2 y = 0,$$

so that the auxiliary equation has two equal roots. Putting

$$(d / dx - m_1)\, y = u, \tag{4.5}$$

above differential equation becomes

$$a_0 (d / dx - m_1)\, u = 0, \qquad \text{i.e.} \qquad du / dx - m_1 u = 0,$$

which has a solution $u = A_2 . e^{m_1 x}$. Accordingly, Eq. (4.5) can be rewritten as

$$dy / dx - m_1 y = A_2 .\, e^{m_1 x}.$$

Being a linear differential equation, by Eq. (2.7), it has a solution

$$y.\, e^{-m_1 x} = A_1 + A_2 . \int e^{m_1 x} . e^{-m_1 x} . dx = A_1 + A_2 . x$$

$\Rightarrow$

$$y = (A_1 + A_2 . x) . e^{m_1 x}$$

giving a complete integral of the differential equation (4.4) whose auxiliary equation has equal roots. Similarly, the nth order differential equation (4.1), whose auxiliary equation has *two* equal roots, has a complete integral

$$y = (A_1 + A_2 . x) . e^{m_1 x} + A_3\, e^{m_3 x} + \ldots + A_n\, e^{m_n x}. \tag{4.6}$$

Generalizing this concept to the case when the auxiliary equation (4.2) has r equal roots a complete integral of Eq. (4.1) is given by

$$y = (A_1 + A_2.\,x + A_3.x^2 + \ldots + A_r.\,x^r).\,e^{m_1 x} + A_{r+1}\; e^{m_{r+1} x} + \ldots + A_n\; e^{m_n x} \tag{4.7}$$

4.3. Complex roots: Let the auxiliary equation have a complex root, say $\alpha + i\beta$. The equation then also has its complex conjugate $\alpha - i\beta$ as another root and $n - 2$ distinct (real) roots $m_3, \ldots, m_n$. Correspondingly, the solution given by Eq. (4.3) becomes

$$y \;=\; A_1 e^{(\alpha + i\beta)x} + A_2\; e^{(\alpha - i\beta)x} + A_3\; e^{m_3 x} + \ldots + A_n\; e^{m_n x}. \tag{4.8a}$$

The sum of first two terms can be simplified as

$$e^{\alpha x}.\{A_1 e^{\,i\beta x} + A_2\; e^{\,-i\beta x}\} \;=\; e^{\alpha x}.\{A_1 (\cos \beta x + i \sin \beta x)$$

$$+ A_2 (\cos \beta x - i \sin \beta x)\} = e^{\alpha x}.\{(A_1 + A_2).\cos \beta x + i\,(A_1 - A_2).\sin \beta x\}$$

$$=\; e^{\alpha x}.(B_1.\cos \beta x + B_2.\sin \beta x) \;=\; C_1.\; e^{\alpha x}.\cos(\beta x + C_2),$$

where we have set $A_1 + A_2 = B_1 = C_1.\cos C_2$ and $i\,(A_1 - A_2) = B_2 = - C_1.\sin C_2$. Hence, Eq. (4.8a) can also be written as

$$y = (B_1.\cos \beta x + B_2.\sin \beta x).\; e^{\alpha x} + A_3\; e^{m_3 x} + \ldots + A_n\; e^{m_n x} \tag{4.8b}$$

$$=\; C_1.\; e^{\alpha x}.\cos(\beta x + C_2) + A_3\; e^{m_3 x} + \ldots + A_n\; e^{m_n x}. \tag{4.8c}$$

4.3.1. Repeated complex roots: If the complex roots $\alpha \pm i\beta$ are repeated the solution in Eq. (4.8a) becomes

$$y = (A_1 + A_2.x).\,e^{(\alpha + i\beta)x} + (A_3 + A_4.\,x).\,e^{(\alpha - i\beta)x} + A_5\,e^{m_5 x} + \ldots + A_n e^{m_n x}$$

$$= \{(B_1 + B_2\,.x).\cos \beta x + (B_3 + B_4.x).\sin \beta x\}.\,e^{\alpha x} + A_5\; e^{m_5 x} + \ldots + A_n\, e^{m_n x}. \tag{4.9}$$

4.4. Pair of roots $\alpha \pm \sqrt{\beta}$: If the auxiliary equation has two roots $\alpha + \sqrt{\beta}$ and $\alpha - \sqrt{\beta}$, the solution in Eq. (4.3) becomes

$$y \;=\; A_1\; e^{(\alpha + \sqrt{\beta})x} + A_2\, e^{(\alpha - \sqrt{\beta})x} + A_3\; e^{m_3 x} + \ldots + A_n\; e^{m_n x}. \tag{4.10a}$$

Analogous to above, the sum of first two terms can be simplified as

$$e^{\alpha x}.\{A_1 e^{x\sqrt{\beta}} + A_2 e^{-x\sqrt{\beta}}\} = e^{\alpha x}[A_1\{\cosh(x\sqrt{\beta}) + \sinh(x\sqrt{\beta})\}$$

$$+ A_2\{\cosh(x\sqrt{\beta}) - \sinh(x\sqrt{\beta})\}]$$

$$= e^{\alpha x}.\{(A_1 + A_2).\cosh(x\sqrt{\beta}) + (A_1 - A_2).\sinh(x\sqrt{\beta})\}$$

$$= e^{\alpha x}.(B_1.\cosh(x\sqrt{\beta}) + B_2.\sinh(x\sqrt{\beta})) = C_1.e^{\alpha x}.\cosh(x\sqrt{\beta} + C_2),$$

where we have put

$$A_1 + A_2 = B_1 = C_1.\cosh C_2 \qquad \text{and} \qquad (A_1 - A_2) = B_2 = C_1.\sinh C_2.$$

Hence, the soln. in Eq. (4.10a) can also be written as

$$y = \{B_1.\cosh(x\sqrt{\beta}) + B_2.\sinh(x\sqrt{\beta})\}.e^{\alpha x} + A_3 e^{m_3 x} + \ldots + A_n e^{m_n x} \qquad (4.10b)$$

$$= C_1.e^{\alpha x}.\cosh(x\sqrt{\beta} + C_2) + A_3 e^{m_3 x} + \ldots + A_n e^{m_n x}. \qquad (4.10c)$$

Example 4.1. Solve the following linear differential equations with constant coefficients, where $D \equiv d/dx$:

(i) $(D^3 + 2D^2 - D - 2)y = 0$, **(ii)** $(D^2 - 8D + 16)y = 0$,

(iii) $(D^3 - 8)y = 0$, **(iv)** $(D^6 + 3D^4 + 3D^2 + 1)y = 0$.

Solution. (i) The auxiliary equation is

$$m^3 + 2m^2 - m - 2 \equiv (m + 2)(m - 1)(m + 1) = 0,$$

having all real distinct roots: $m = 1, -1$ and -2. So, the solution of the differential equation, by Eq. (4.3), is

$$y = A_1 e^{x} + A_2 e^{-x} + A_3 e^{2x}.$$

(ii) The auxiliary equation is

$$m^2 - 8m + 16 \equiv (m - 4)^2 = 0,$$

with repeated real roots: $m = 4, 4$. So, the solution of the differential equation, by Eq. (4.6), is

$$y = (A_1 + A_2 x)e^{4x}.$$

(iii) The auxiliary equation is

$$m^3 - 8 \equiv (m-2)(m^2+2m+4) = 0$$

with one real and two complex roots: $m = 2, -1 \pm \sqrt{3}\, i$. So, the solution of the differential equation, by Eqs. (4.3) and (4.8b), is

$$y = A_1 e^{2x} + (A_2 \cos \sqrt{3}\, x + A_3 \sin \sqrt{3}\, x).\, e^{-x}.$$

(iv) The auxiliary equation is

$$m^6 + 3m^4 + 3m^2 + 1 = \{(m^2)^3 + 1\} + 3m^2(m^2+1)$$

$$= (m^2+1)(m^4 - m^2 + 1) + 3m^2(m^2+1)$$

$$= (m^2+1)(m^4 + 2m^2 + 1) = (m^2+1)(m^2+1)^2 = (m^2+1)^3 = 0$$

with (thrice) repeated complex roots: $m = \pm\, i$. So, the solution of the differential equation, by Eq. (4.9), is

$$y = (A_1 + A_2 x + A_3 x^2) \cos x + (A_4 + A_5 x + A_6 x^2) \sin x. \ //$$

§ 5. The Particular Integral

We return to the discussion of integration of the differential equation of the type given by Eq. (1.4), where the RHS does not vanish any more. Treating the LHS of the equation as some function, say $f(D)$ of the differential operator $D \equiv d/dx$ operated upon y, a more general form of the differential equation can be considered as:

$$f(D)\, y = R(x). \qquad (5.1)$$

The *particular integral* of the differential equation is defined as

$\{1/f(D)\}\, R(x)$ also written as $\{f(D)\}^{-1} R(x)$.

The operators $f(D)$ and $1/f(D) \equiv \{f(D)\}^{-1}$ are inverse to each other, i.e.

$$f(D)\, [\{f(D)\}^{-1} R(x)] = R(x). \qquad (5.2)$$

In particular,

$$D\{D^{-1}R(x)\} = R(x) \quad \Rightarrow \quad D^{-1}R(x) = \int R(x)\,dx. \quad (5.3)$$

Similarly, we may evaluate $(D-a)^{-1}R(x)$, where a is a constant.

Theorem 5.1. $\quad (D-a)^{-1}R(x) = e^{ax}.\int R(x).\,e^{-ax}.\,dx. \quad (5.4)$

Proof. If $(D-a)^{-1}R(x) = u(x)$ then, by definition, there results a linear differential equation

$$(D-a)\,u \equiv du/dx - a\,u = R(x)$$

having a solution

$$u(x).\,e^{-ax} = \int R(x).\,e^{-ax}.\,dx,$$

that gives the desired result.

In case the operator function $f(D)$ can be factorized into linear factors:

$$f(D) = (D-a_1)(D-a_2)\ldots(D-a_n),$$

above Theorem helps to evaluate $\{f(D)\}^{-1}R(x)$. //

Theorem 5.2. We have

$$\{f(D)\}^{-1}R = A_1.e^{a_1x}.\int R.e^{-a_1x}.\,dx + A_2.e^{a_2x}.\int R.e^{-a_2x}.\,dx$$

$$+ \ldots + A_n.e^{a_nx}.\int R.e^{-a_nx}.\,dx. \quad (5.5)$$

Proof. Breaking $1/f(D)$ into partial fractions:

$$1/f(D) = A_1/(D-a_1) + A_2/(D-a_2) + \ldots + A_n/(D-a_n),$$

and applying Eq. (5.4), we derive Eq. (5.5). //

Note 5.1. In case $R(x) = x^m$, $\{f(D)\}^{-1}x^m$ can be evaluated by expanding $\{f(D)\}^{-1}$ in ascending powers of D and the expansion so obtained operates on the function x^m.

Theorem 5.3. For any constant a satisfying $f(a) \neq 0$, there holds

$$\{1/f(D)\}\,e^{ax} = \{1/f(a)\}\,e^{ax} \quad (5.6)$$

Proof. Expanding $f(D)$ in powers of D:

$$f(D) = a_0 . D^n + a_1 . D^{n-1} + \ldots + a_{n-1} . D + a_n,$$

so that

$$f(D)e^{ax} = (a_0 . D^n + a_1 . D^{n-1} + \ldots + a_{n-1} . D + a_n)e^{ax}$$

$$= (a_0 . a^n + a_1 . a^{n-1} + \ldots + a_{n-1} . a + a_n)e^{ax} = f(a) . e^{ax}, \quad (5.7)$$

for

$$De^{ax} = a . e^{ax}, \; D^2 e^{ax} = a^2 . e^{ax}, \; \ldots, \; D^n e^{ax} = a^n . e^{ax}.$$

Therefore,

$$f(D) \{e^{ax} / f(a)\} = e^{ax} \quad \Rightarrow \text{the result. //}$$

Corollary 5.1. For any constant a satisfying $f(a) \neq 0$, there hold

$$\left. \begin{aligned} \{1/f(D^2)\} \cos ax &= \{1/f(-a^2)\} \cos ax, \\ \{1/f(D^2)\} \sin ax &= \{1/f(-a^2)\} \sin ax. \end{aligned} \right\} \quad (5.8)$$

Proof. Replacing D by D^2 and a by ai in Eq. (5.6), we obtain

$$\{1/f(D^2)\} e^{axi} = e^{axi} / f\{(ai)^2\} = e^{axi} / f(-a^2),$$

or, by Euler's theorem

$$\{1/f(D^2)\} (\cos ax + i \sin ax) = (\cos ax + i \sin ax) / f(-a^2).$$

Equating real and imaginary parts on either side we get the desired results. //

Note 5.2. If $f(D)$ also contains the odd powers of D, to operate $\cos ax$, and $\sin ax$ by $1/f(D)$, we proceed as follows:

Separating the terms containing odd and even powers of D in $f(D)$:

$$f(D) = f_1(D^2) + D . f_2(D^2),$$

we obtain,

$$\{1/f(D)\} \cos ax = [1/\{f_1(D^2) + D . f_2(D^2)\}] \cos ax$$

$$= [1/\{f_1(-a^2) + D . f_2(-a^2)\}] \cos ax, \text{ by Eqs. (5.8).}$$

Putting $f_1(-a^2) = a_1$, $f_2(-a^2) = a_2$ and introducing the operator

$a_1 - a_2 D$ in numerator as well as in the denominator the RHS of above relation reduces to

$$(a_1 - a_2 D)\,[\{1 / (a_1 - a_2 D)\,(a_1 + a_2 D)\}\cos ax]$$

$$= (a_1 - a_2 D)\,[\{1 / (a_1^2 - a_2^2 D^2)\}\cos ax]$$

$$= (a_1 - a_2 D)\,[\{1 / (a_1^2 + a_2^2 . a^2)\}\cos ax], \text{ again by Eqs. (5.8)}$$

$$= \{a_1 . \cos ax - a_2 . D(\cos ax)\} / (a_1^2 + a^2 a_2^2)$$

$$= \{a_1 . \cos ax + a.a_2 . \sin ax)\} / (a_1^2 + a^2 a_2^2). \qquad (5.9)$$

Similarly, we may derive

$$\{1 / f(D)\}\sin ax = \{a_1 . \sin ax - a.a_2 . \cos ax)\} / (a_1^2 + a^2 . a_2^2). \qquad (5.10)$$

Theorem 5.4. For a non-zero constant a, there hold

$$\left.\begin{aligned} \{1 / (D^2 + a^2)\}\cos ax &= (x / 2a)\sin ax, \\ \{1 / (D^2 + a^2)\}\sin ax &= -(x / 2a)\cos ax. \end{aligned}\right\} \qquad (5.11)$$

Proof. Since $f(-a^2) = 0$ for the present operator function $f(D) \equiv D^2 + a^2$, the results of Corollary 5.1 cannot be applied. Factorizing $D^2 + a^2$ into linear factors $D + ai$ and $D - ai$ and breaking $1 / (D^2 + a^2)$ into partial fractions:

$$1 / (D^2 + a^2) = 1 / (D + ai)(D - ai) = (1/2ai)\{1 / (D - ai) - 1 / (D + ai)\},$$

we evaluate

$$\{1 / (D^2 + a^2)\}\cos ax = (1/2ai)\{1/(D - ai) - 1/(D + ai)\}\cos ax. \qquad (5.12)$$

But,

$$\{1 / (D - ai)\}\cos ax = e^{axi} . \int e^{-axi} . \cos ax . dx, \qquad \text{by Eq. (5.4)}$$

$$= (e^{axi} / 2)\int e^{-axi} . (e^{axi} + e^{-axi})\,dx = (e^{axi} / 2)\int (1 + e^{-2axi})\,dx$$

$$= (e^{axi} / 2)\,(x - e^{-2axi} / 2ai).$$

Similarly,

$$\{1 / (D + ai)\}\cos ax = (e^{-axi} / 2)\,(x + e^{2axi} / 2ai).$$

Hence, Eq. (5.12) reduces to

$$\{1 / (D^2 + a^2)\} \cos ax = (1/2ai) \{(e^{axi} / 2) (x - e^{-2ax\,i} / 2ai)$$

$$- (e^{-axi} /2) (x + e^{2axi} / 2ai)\} = (1/2a) \{x. \sin ax + (\cos ax)/2a\}. \quad (5.13)$$

Since the *complementary function* in the solution of the differential equation

$$(D^2 + a^2)\, y = \cos ax,$$

by Eq. (4.8b), is $B_1. \cos ax + B_2. \sin ax$ the term $(\cos ax) / 4a^2$ in Eq. (5.13) may be dropped. Hence, Eq. (5.13) reduces to the desired form.

Similarly, the other result can be established. //

Theorem 5.5. $\{1/f(D)\}\{e^{ax}. u(x)\} = e^{ax}.\{1/f(D+a)\}\, u(x)$. (5.14)

Proof. We note that

$$D\,\{e^{ax}. u\} = e^{ax}. D\,u + a. e^{ax}. u = e^{ax} (D + a)\, u,$$

$$D^2 \{e^{ax}. u\} = D\,\{e^{ax} (D + a)\, u\}$$

$$= a. e^{ax} (D + a)\, u + e^{ax}. D\,(D + a)\, u = e^{ax}. (D + a)^2\, u,$$

and in general (by successive differentiation):

$$D^n \{e^{ax}. u\} = e^{ax}. (D + a)^n\, u.$$

Thus,

$$f(D)\,\{e^{ax}. u(x)\} = e^{ax}. \{f(D + a)\, u(x)\}.$$

Replacing the function $u(x)$ by $\{1/f(D + a)\}\, u(x)$ in above relation we also get

$$f(D)\,[e^{ax}.\{1/f(D + a)\}\, u] = e^{ax}. f(D + a)\{1/f(D + a)\}u = e^{ax}.u(x)$$

implying the desired result. //

Note 5.3. When $f(a) = 0$, $\{1/f(D)\}\, e^{ax}$ may be evaluated by Theorem 5.5 by taking $u(x) = 1$.

Theorem 5.6.

$$\{1/f(D)\}\{x.\,u(x)\} = x.\{1/f(D)\}u(x) + [d\{f(D)\}^{-1}/dD]\,u(x)$$

$$= x.\{1/f(D)\}u(x) - [f'(D)/\{f(D)\}^2]\,u(x), \qquad (5.15)$$

prime denoting derivation w.r.t. D.

Proof. We note that

$$D(x\,u) = x\,D\,u + u, \quad D^2(x\,u) = x\,D^2\,u + 2\,Du,$$

and, similarly,

$$D^n(x\,u) = x\,D^n\,u + n\,D^{n-1}\,u = x\,D^n\,u + \{d\,(D^n)/dD\}u.$$

Thus, in general,

$$f(D)\,(x\,u) = x\,\{f(D)\,u\} + [d\{f(D)\}/dD]\,u. \qquad (5.16)$$

Replacing u by $\{1/f(D)\}\,u$, above equation also yields

$$f(D)\,[x\{1/f(D)\}u] = x f(D)\,\{1/f(D)\}u + [d\{f(D)\}/dD]\{1/f(D)\}u$$

$$= x.\,u + \{f'(D)/f(D)\}\,u.$$

Further, operating it by $1/f(D)$ gives

$$x\,\{1/f(D)\}\,u = \{1/f(D)\}\,(x\,u) + [f'(D)/\{f(D)\}^2]\,u,$$

which is same as Eq. (5.15). //

Example 5.1. Solve the following linear differential equations:

(i) $(D^2 + 2D + 10)\,y + 37\sin 3x = 0$, when $y = 3$, $dy/dx = 0$ at $x = 0$;

(ii) $(D^2 + a^2)y = \sec ax$, **(iii)** $(D^3 + 3D^2 + 2D)y = x^2$,

(iv) $(D-1)^2(D^2+1)^2 y = \sin x$, **(v)** $(D^2 + 3D + 2)y = \cos 2x$,

(vi) $(D^2 - 6D + 13)\,y = 8\,e^{3x}.\sin 2x$,

(vii) $(D^2 - 4D + 4)y = 8x^2.\,e^{2x}.\sin 2x$.

Solution. (i) The auxiliary equation is $m^2 + 2m + 10 = 0$ having distinct complex roots $m = -1 \pm 3i$. Hence, by Eq. (4.8b), the complementary function is

$$\text{C.F.} = (A_1 \cos 3x + A_2 \sin 3x)\, e^{-x}.$$

Also, the particular integral is

$$\text{P.I.} = (D^2 + 2D + 10)^{-1} (-37.\sin 3x)$$

$$= -37.(-9 + 2D + 10)^{-1} (\sin 3x), \qquad \text{by Eqs. (5.8)}$$

$$= -(37/2)(D + 1/2)^{-1} (\sin 3x) = -(37/2)\, e^{-x/2} \int e^{x/2}.\sin 3x.\, dx,$$

by Eq. (5.4). Applying the integral formula

$$\int e^{ax}.\sin bx.\, dx = e^{ax}.(a.\sin bx - b.\cos bx) / (a^2 + b^2), \qquad (5.17)$$

the particular integral is evaluated as

$$\text{P.I.} = -(37/2)\, \{(1/2)\sin 3x - 3\cos 3x) / (1/4 + 9) = 6\cos 3x - \sin 3x.$$

Hence, the solution of the differential equation is

$$y = \text{C.F.} + \text{P.I.} = (A_1 \cos 3x + A_2 \sin 3x)\, e^{-x} + 6\cos 3x - \sin 3x \qquad (5.18)$$

$$\Rightarrow \qquad dy / dx = \{-A_1 (\cos 3x + 3\sin 3x) + A_2 (3\cos 3x - \sin 3x)\}\, e^{-x}$$

$$- 3\,(6 \sin 3x + \cos 3x).$$

Applying the initial conditions, we derive $A_1 = -3$ and $A_2 = 0$; which reduce Eq. (5.18) as

$$y = (6 - 3e^{-x}).\cos 3x - \sin 3x.$$

(ii) The auxiliary equation $m^2 + a^2 = 0$ has distinct imaginary roots: $m = \pm\, a\, i$. Hence, by Eq. (4.8b), the complementary function is

$$\text{C.F.} = A_1 \cos ax + A_2 \sin ax.$$

Also, the particular integral is

$$\text{P.I.} = \{1 / (D^2 + a^2)\} \sec ax = \{1 / (D + ai)(D - ai)\} \sec ax$$

$$= (1/2ai)\,\{(D - ai)^{-1} - (D + ai)^{-1}\}\ \sec ax.$$

By Eq. (5.4),

$$(D - ai)^{-1}\ \sec ax\ =\ e^{axi}.\int e^{-axi}.\ \sec ax.\ dx$$

$$= e^{axi}.\int \{(\cos ax - i.\ \sin ax)\,/\,\cos ax\}\ dx$$

$$=\ e^{axi}.\int (1 - i.\ \tan ax)\ dx\ =\ e^{axi}.\{x + (i\,/a).\ \ln \cos ax\}.$$

Similarly,

$$(D + ai)^{-1}\ \sec ax\ =\ e^{-axi}.\ \{x - (i\,/a).\ \ln \cos ax\}.$$

Hence,

$$\text{P.I.}\ =\ x\,(e^{axi} - e^{-axi})\,/\,2ai + (\ln \cos ax)\,(e^{axi} + e^{-axi})\,/\,2a^2$$

$$=\ (x\,/a)\ \sin ax + (1/a^2)\ \cos ax.\ (\ln \cos ax),$$

and the complete solution is

$$y\ =\ \text{C.F.} + \text{P.I.}\ =\ A_1 \cos ax + A_2 \sin ax + (x\,/a)\ \sin ax$$

$$+ (1/\,a^2)\ \cos ax.\ (\ln \cos ax).$$

(iii) The auxiliary equation

$$m^3 + 3m^2 + 2m\ \equiv\ m\,(m + 1)\,(m + 2) =\ 0$$

has all distinct roots: $m = 0, -1$ and 2. Hence, by Eq. (4.3), we have

$$\text{C.F.}\ =\ A_1 + A_2\,e^{-x} + A_3\,e^{-2x}.$$

Since

$$1/\,(D^3 + 3D^2 + 2D) = (1/2D)\,(1 + 3D/2 + D^2/2)^{-1}$$

$$= (1/2D)\,\{1 - (3D/2 + D^2/2) + (3D/2 + D^2/2)^2 - (3D/2 + D^2/2)^3 + \ldots\}$$

$$= 1/2D - 3/4 + 7D/8 - 15D^2/16 - \ldots$$

$$\text{P.I.} = \{1/\,(D^3 + 3D^2 + 2D)\}(x^2) = (1/2D - 3/4 + 7D/8 - 15D^2/16 - \ldots)\,x^2$$

$$=\ (1/2)\int x^2\,dx - 3x^2/4 + 7x/4 - 15/8 = x^3/6 - 3x^2/4 + 7x/4 - 15/8.$$

Dropping the constant term in the P.I. as it is already included within A_1 in the C.F., the complete solution is

$$y = \text{C.F.} + \text{P.I.} = A_1 + A_2 e^{-x} + A_3 e^{-2x} + (2x^3 - 9x^2 + 21x) / 12.$$

(iv) The auxiliary equation $(m-1)^2 (m^2+1)^2 = 0$ has one real root $m = 1$ and two imaginary roots $m = \pm i$ (all repeating once). Hence, by Eqs. (4.6) and (4.9), we have

$$\text{C.F.} = (A_1 + A_2 x) e^{x} + (A_3 + A_4 x) \cos x + (A_5 + A_6 x) \sin x.$$

Also, by Eq. (5.8)

$$\text{P.I.} = \{1/(D-1)^2 (D^2+1)^2\} \sin x = \{1/(-1-2D+1)(D^2+1)^2\} \sin x$$

$$= \{1/2(D^2+1)^2\} (1/D)(-\sin x) = \{1/2(D^2+1)^2\} \int (-\sin x)\, dx$$

$$= \{1/2(D^2+1)^2\} \cos x = \{1/4(D^2+1)\} (x. \sin x), \quad \text{by Eq. (5.11)}$$

$$= \{1/4(D^2+1)\} \{\text{Im}(x. e^{ix})\} = (1/4) \text{Im} [\{1/(D^2+1)\}(x. e^{ix})\}$$

$$= (1/4) \text{Im} [e^{ix}. \{(D+i)^2 + 1\}^{-1} x] = (1/4) \text{Im} \{e^{ix}.(D^2 + 2iD)^{-1} x\},$$

by Eq. (5.14). But,

$$1/(D^2 + 2iD) = (1/2iD)(1 + D/2i)^{-1} = (1/2iD)(1 - D/2i - D^2/4 + \ldots)$$

$$= 1/2iD + 1/4 - D/8i + \ldots$$

$\Rightarrow$

$$(D^2 + 2iD)^{-1} x = (1/2i) \int x\, dx + x/4 - 1/8i = x^2/4i + x/4 - 1/8i$$

$$= i(1 - 2x^2)/8 + x/4.$$

Therefore,

$$\text{P.I.} = (1/4) \text{Im} [(\cos x + i \sin x).\{i(1-2x^2)/8 + x/4\}]$$

$$= (1/32) \{(1 - 2x^2) \cos x + 2x. \sin x\}.$$

The terms containing $\cos x$ and $x. \sin x$ in the P.I. are already included in the C.F. so may be dropped. Hence, the complete solution of the differential equation is

$$y = \text{C.F.} + \text{P.I.}$$

$$= (A_1 + A_2 x) e^{x} + (A_3 + A_4 x - x^2/16) \cos x + (A_5 + A_6 x) \sin x.$$

(v) The auxiliary equation $m^2 + 3m + 2 \equiv (m + 1)(m + 2) = 0$ has distinct real roots $m = -1, -2$. Hence, by Eq. (4.3), we have

$$\text{C.F.} = A_1 e^{-x} + A_2 e^{-2x}.$$

Also, by Eq. (5.8)

$$\text{P.I.} = \{1 / (D^2 + 3D + 2)\} \cos 2x = \{1 / (-2^2 + 3D + 2)\} \cos 2x$$

$$= (1/3)(D - 2/3)^{-1} \cos 2x = (1/3)\, e^{2x/3} \int e^{-2x/3} \cos 2x\, dx, \text{ by Eq. (5.4)}$$

or, on application of the formula

$$\int e^{ax}. \cos bx. dx = e^{ax}.(a. \cos bx + b. \sin bx) / (a^2 + b^2), \quad (5.19)$$

$$\text{P.I.} = (1/3)\, \{-(2/3) \cos 2x + 2 \sin 2x)\} / (4/9 + 4)$$

$$= (3 \sin 2x - \cos 2x) / 20.$$

Hence, the solution is

$$y = \text{C.F.} + \text{P.I.} = A_1 e^{-x} + A_2 e^{-2x} + (3 \sin 2x - \cos 2x) / 20.$$

(vi) The auxiliary equation is $m^2 - 6m + 13 = 0$ having complex roots $m = 3 \pm 2i$. Hence, by Eq. (4.8b), the complementary function is

$$\text{C.F.} = (A_1 \cos 2x + A_2 \sin 2x)\, e^{3x}.$$

Also, by Eq. (5.14), we have

$$\text{P.I.} = \{1/ (D^2 - 6D + 13)\}\, (8e^{3x}. \sin 2x)$$

$$= 8e^{3x}. \{(D + 3)^2 - 6\, (D + 3) + 13\}^{-1} (\sin 2x)$$

$$= 8e^{3x}. (D^2 + 4)^{-1} (\sin 2x) = 8e^{3x}.(-x / 4) \cos 2x = -2x. e^{3x}. \cos 2x,$$

by (5.11). Hence, the solution is

$$y = \text{C.F.} + \text{P.I.} = \{(A_1 - 2x) \cos 2x + A_2 \sin 2x)\}e^{3x}.$$

(vii) The auxiliary equation $m^2 - 4m + 4 \equiv (m - 2)^2 = 0$ has equal roots $m = 2, 2$. Hence, by Eq. (4.6), we have

$$\text{C.F.} = (A_1 + A_2 x)\, e^{2x}.$$

Also, by Eq. (5.14),

$$\text{P.I.} = \{1/(D-2)^2\}(8x^2.e^{2x}.\sin 2x) = 8e^{2x}.\{1/(D+2-2)^2\ (x^2.\sin 2x)$$

$$= 8e^{2x}.\ (1/D^2)\ \{\text{Im}\ (x^2.\ e^{2ix})\} = 8e^{2x}.\text{Im}\ \{(1/D^2)\ (x^2.\ e^{2ix})\}$$

$$= 8e^{2x}.\ \text{Im}\ [e^{2ix}\ \{1/(D+2i)^2\}\ x^2],$$

by Eq. (5.14). But,

$$1/(D+2i)^2 = (1/2i)^{-2}/(1+D/2i)^2 = -(1/4)\ (1-iD/2)^{-2}$$

$$= -(1/4)\ (1+iD-3D^2/4-iD^3/2+\ldots)$$

$\Rightarrow$

$$1/(D+2i)^2\}\ x^2 = -(1/4)\ (x^2+2ix-3/2).$$

Therefore,

$$\text{P.I.} = -2e^{2x}.\ \text{Im}\ \{(\cos 2x+i\sin 2x)\ (x^2+2ix-3/2)\}$$

$$= -2\ e^{2x}\ \{2x.\ \cos 2x+(x^2-3/2)\sin 2x\},$$

and the solution is

$$y = \text{C.F.}+\text{P.I.} = \{A_1+A_2x-4x.\ \cos 2x+(3-2x^2)\sin 2x\}e^{2x}.\ //$$

§ 6. Homogeneous Linear Differential Equations of Any Order

A differential equation of the form

$$(x^n.\ D^n + a_1.x^{n-1}.\ D^{n-1} + a_2.\ x^{n-2}.\ D^{n-2} + \ldots$$

$$+ a_{n-1}.x.\ D + a_n)y = R\,(x) \qquad (6.1)$$

is called a *homogeneous linear differential equation*. Applying the change of independent variable:

$$x = e^t \quad \Rightarrow dx/dt = e^t \Rightarrow dt/dx = 1/e^t = 1/x, \quad (6.2)$$

and therefore

$$D \equiv d/dx = (dt/dx)\ (d/dt) = (1/x).\ d/dt \Rightarrow x.\ D = d/dt \equiv D'. \quad (6.3)$$

Operating the equation (6.3) by itself, there also results

$$x.D\,(x.\,D) = (d/dt)\,(d/dt), \text{ i.e. } x^2.D^2 + x.\,D = (d/dt)\,(d/dt)$$

$\Rightarrow$

$$x^2.\,D^2 = (d/dt)\,(d/dt) - x.\,D = (d/dt)\,(d/dt - 1) = D'\,(D' - 1), \quad (6.4)$$

by Eq. (6.3). Operating it again by $x.\,D \equiv D'$, we derive

$$x^3.D^3 + 2x^2.\,D^2 = D'\,\{D'\,(D' - 1)\}$$

$\Rightarrow$

$$x^3.\,D^3 = D'\,\{D'\,(D' - 1)\} - 2D'\,(D' - 1)\} = (D' - 2)\,\{D'\,(D' - 1)\}$$

$$= D'\,(D' - 1)\,(D' - 2), \quad (6.5)$$

for Eq. (6.4) and the commutative properties of the operators D', $D' - 1$ and $D' - 2$.

Similarly, by the method of finite induction, we may also derive

$$x^n.\,D^n = D'\,(D' - 1)\,(D' - 2)\,\ldots\,\{D' - (n - 1)\}. \quad (6.6)$$

Putting from Eqs. (6.3) – (6.6), the equation (6.1), reduces to a linear form with constant coefficients discussed in the preceding Section:

$$f(D')\,y = R\,(x). \quad (6.7)$$

6.1. Complementary function: Setting $y = x^m$ so that

$$D\,y = m.\,x^{m-1}, \qquad D^2\,y = m.\,(m - 1).\,x^{m-2}, \ldots,$$

$$D^n\,y = m.\,(m - 1)\,(m - 2)\,\ldots\,(m - n + 1)\,x^{m-n}$$

$\Rightarrow$

$$x.\,D\,y = m\,x^m, \qquad x^2D^2\,y = m.\,(m - 1)\,x^m, \ldots,$$

$$x^n.\,D^n\,y = m.\,(m - 1).\,(m - 2)\,\ldots\,(m - n + 1)\,x^m,$$

and substituting for all above in the LHS of Eq. (6.1), we derive the auxiliary equation

$$\{\,m\,(m - 1)\,(m - 2)\,\ldots\,(m - n + 1) + m\,(m - 1)\,(m - 2)\,\ldots\,(m - n + 2)\,a_1$$

$$+ \ldots + m\,a_{n-1} + a_n\}\,x^m = 0, \quad (6.8)$$

which is of nth degree in m. For n distinct roots $m_1, m_2, \ldots, m_n$ the complementary function of the solution of Eq. (6.1) is

$$\text{C.F.} = A_1 x^{m_1} + A_2\, x^{m_2} + \ldots + A_n\, x^{m_n}\,. \tag{6.9}$$

6.2. Particular integral: In analogy with the discussion in the preceding section, the particular integral of Eq. (6.7) can be defined by

$$\{f(D')\}^{-1} R \;\equiv\; \{1/f(D')\}\, R.$$

Theorem 6.1.

$$(D' - a)^{-1} R(x) = x^{a}. \int R(x).\, x^{-(a+1)}.\, dx. \tag{6.10}$$

Proof. Taking $(D' - a)^{-1} R(x) = u(x)$, there results a differential equation

$$(D' - a)\, u \equiv x.\, (du/dx) - a\, u = R(x), \quad \text{or,} \quad du/dx - (a/x).\, u = R/x,$$

which is linear in u and du/dx having integrating factor

$$e^{-\int (a/x)\, dx} = e^{-a.\ln x} = x^{-a}.$$

Hence, its solution is

$$u(x).\, x^{-a} = \int (R/x).\, x^{-a}.\, dx = \int R(x).\, x^{-(a+1)}.\, dx. \;//$$

Note 6.1. The result can also be derived directly from Eq. (5.4). With the change of variables as per Eq. (6.2) and $D' \equiv d/dt$, the result in Eq. (5.4) is rewritten as

$$(D' - a)^{-1} R(e^{t}) = e^{at}. \int R(e^{t}).\, e^{-at}.\, dt = x^{a}. \int R(x).\, x^{-a}.\, (1/x).\, dx$$

assuming the form given by Eq. (6.10).

The other cases discussed in the preceding Section can be similarly established for the differential equation (6.7).

Example 6.1. Solve the following homogeneous linear differential equations with variable coefficients:

(i) $(x^2.D^2 - x.\, D - 3)\, y = x^2.\, \ln x$, **(ii)** $(x^2.\, D^2 + 4x.\, D + 2)\, y = e^{x}$.

Solution. (i) Applying the change of parameters given by Eq. (6.2), and putting from Eqs. (6.3) and (6.4), the differential equation reduces to

$$\{D'(D'-1)-D'-3\}\,y \equiv (D'^2-2D'-3)\,y = t.\,e^{2t}.$$

Its auxiliary equation $m^2 - 2m - 3 = 0$ has distinct roots $m = -1, 3$. Hence, by Eq. (4.3), we get

$$\text{C.F.} = A_1\,e^{-t} + A_2\,e^{3t} = A_1\,x^{-1} + A_2\,x^3.$$

Also, by Eq. (5.14),

$$\text{P.I.} = \{1/(D'^2-2D'-3)\}(t.\,e^{2t}) = e^{2t}\,\{1/(D'+2)^2-2\,(D'+2)-3\}t$$

$$= e^{2t}\,\{1/(D'^2+2D'-3)\,t = -(e^{2t}/3)\,\{1-(2D'+D'^2)/3\}^{-1}\,t$$

$$= -(e^{2t}/3)\,\{1+(2D'+D'^2)/3+\ldots\}t = -(e^{2t}/3)\,(t+2/3)$$

$$= -(x^2/3)\,\{\ln x + 2/3).$$

Hence, the solution is

$$y = \text{C.F.} + \text{P.I.} = A_1\,x^{-1} + A_2\,x^3 - (x^2/3)\,\{\ln x + 2/3).$$

(ii) As above, under the change of parameters given by Eq. (6.2), the differential equation transforms to

$$\{D'(D'-1)+4D'+2\}y \equiv (D'^2+3D'+2)\,y = e^{x}.$$

Its auxiliary equation $m^2 + 3m + 2 = 0$ has distinct roots $m = -1, -2$. Hence, by Eq. (4.3), we get

$$\text{C.F.} = A_1\,e^{-t} + A_2\,e^{-2t} = A_1\,x^{-1} + A_2\,x^{-2}.$$

Breaking into partial fractions

$$1/(D'^2+3D'+2) = 1/(D'+1)(D'+2) = 1/(D'+1) - 1/(D'+2),$$

and applying Eq. (6.10), we have

$$\text{P.I.} = \{1/(D'+1) - 1/(D'+2)\}e^x = x^{-1}\int e^x\,.x^{1-1}.\,dx - x^{-2}\int e^x.x^{2-1}.\,dx$$

$$= x^{-1}.e^x - x^{-2}(x.\,e^x - \int e^x.\,dx) = x^{-1}.e^x - x^{-1}.e^x + x^{-2}.e^x = x^{-2}.\,e^x.$$

Hence, the solution is

$$y = \text{C.F.} + \text{P.I.} = A_1 / x + (A_2 + e^{x}) / x^2.$$

§ 7. Differential Equations Reducible to Homogeneous Linear Form

A differential equation

$$\{ (ax + b)^n . D^n + a_1 . (ax + b)^{n-1} . D^{n-1} + \ldots$$

$$+ a_{n-1} .(ax + b). D + a_n \} y = R(x), \qquad (7.1)$$

with a, b, a_1, ... , a_n as constants, can be reduced to a homogeneous linear form with constant coefficients. Applying the change of variable: $ax + b = z$ so that

$$dz/dx = a, \; D y \equiv dy/dx = (dy/dz).(dz/dx) = a\,(dy/dz) \equiv a.D' y,$$

$$D^2 y \equiv (d / dx)\,(dy / dx) = (d / dx)\,\{a\,(dy / dz)\}$$

$$= a\,(d^2 y / dz^2)\,(dz / dx) = a^2 (d^2 y / dz^2) = a^2 . D'^2 y.$$

Similarly, $D^n y = a^n . D'^n y$. Substitution for these derivatives reduces the equation (7.1) to

$$\{ (a.z)^n . D'^n + a_1 . (az)^{n-1} . D'^{n-1} + \ldots + a_{n-1} . (az). D' + a_n \} y$$

$$= R\,\{(z - b) / a\},$$

or, on division by a^n,

$$\{ z^n . D'^n + (a_1/a). z^{n-1} . D'^{n-1} + \ldots + (a_{n-1} / a^{n-1}). z. D'$$

$$+ (a_n / a^n) \} y = (1/ a^n). R\,\{(z - b) / a\},$$

which is a homogeneous linear differential equation in the independent variable z and dependent variable y.

Example 7.1. Solve the following linear differential equations with variable coefficients:

$$(1 + 2x)^2 . D^2 - 6\,(1 + 2x). D + 16)\, y = 8\,(1 + 2x)^2,$$

where $y = 0$ and $dy / dx = 2$ at $x = 0$.

Solution. Writing $z = 1 + 2x$ so that $dz/dx = 2$ and

$$Dy \equiv dy/dx = (dy/dz)\,(dz/dx) = 2\,dy/dz \equiv 2D'\,y \text{ (say)}.$$

Operating it again by $D \equiv dy/dx$, we get

$$D^2 y = (d/dx)\,(dy/dx) = (d/dx)\,(2dy/dz) = 2\,(d^2 y/dz^2)\,(dz/dx) = 4(D'^2 y).$$

Putting for these derivatives, the given differential equation transforms to

$$(z^2.\,D'^2 - 3z.\,D' + 4)\,y = 2z^2. \tag{7.2}$$

This is a homogeneous linear differential equation (in dependent variable y and its derivatives) with variable coefficients and can now be solved by methods of § 6. Thus, applying the change of parameters

$$z = e^t \quad \Rightarrow \quad dz / dt = e^t \quad \Rightarrow \quad dt / dz = 1/ e^t = 1/z,$$

and

$$D' \equiv d / dz = (dt / dz)\,(d / dt) = (1/z).\, d / dt$$

$\Rightarrow$

$$z.\,D' = d / dt \equiv D'' \text{ (say)}. \tag{7.3}$$

Operating it again by $z.D'$ i.e. D'', there results

$$z.\,D'\,(z.\,D') \equiv z.\,D' + z^2.\,D'^2 = D''\,(D'') = D''^2$$

$\Rightarrow$

$$z^2.\,D'^2 = D''^2 - z.\,D' = D''^2 - D'' = D''\,(D'' - 1). \tag{7.4}$$

Hence, the differential equation (7.2) further transforms to

$$\{D''\,(D'' - 1) - 3D'' + 4\}y \equiv (D''^2 - 4D'' + 4)\,y = 2\,e^{2t}. \tag{7.5}$$

Its auxiliary equation $m^2 - 4m + 4 = 0$ has repeated roots $m = 2, 2$. Hence, by Eq. (4.6), we get

$$\text{C.F.} = (A_1 + A_2\,t)\,e^{2t} = (A_1 + A_2 \ln z)\,z^2$$

$$= \{A_1 + A_2 \ln (1 + 2x)\}\,(1 + 2x)^2.$$

Since the condition of Theorem 5.3 presently does not hold, the P.I. of the differential equation (7.5) can be evaluated by Theorem, 5.5:

$$\text{P.I.} = \{1/(D'' - 2)^2\}(2e^{2t}) = 2e^{2t}.\{1/D''^2\}(1) = 2e^{2t}.\, t^2/2 = t^2.\, e^{2t}$$

$$= z^2.\,(\ln z)^2 = (1+2x)^2.\,\{\ln(1+2x)\}^2.$$

Hence, the solution is

$$y = \text{C.F.} + \text{P.I.} = [A_1 + A_2 \ln(1+2x) + \{\ln(1+2x)\}^2](1+2x)^2 \quad (7.6)$$

$\Rightarrow$

$$dy/dx = 4\,[A_1 + A_2 \ln(1+2x) + \{\ln(1+2x)\}^2](1+2x)$$

$$+ \{2A_2 + 4.\ln(1+2x)\}(1+2x).$$

Applying the initial conditions we evaluate $A_1 = 0$ and $A_2 = 1$. Hence, the solution (7.6) reduces to

$$y = (1+2x)^2.\,\{1 + \ln(1+2x)\}.\,\ln(1+2x). \; //$$

§ 8. Simultaneous Differential Equations

The preceding sections dealt with the solutions of a single differential equation involving one independent variable and one dependent variable. In dynamics, sometimes we encounter with a system of equations involving some variables and the derivatives of independent variables. The simplest class of such equations is the one containing one independent variable and other variable(s) as its functions.

8.1. Simultaneous linear differential equations with constant coefficients: Let t be an independent variable and $x(t)$ and $y(t)$ be two dependent variables. We consider the simultaneous equations

$$f_1(D)\,x(t) + g_1(D)\,y(t) = R_1(t), \quad f_2(D)\,x(t) + g_2(D)\,y(t) = R_2(t), \quad (8.1)$$

where $D \equiv d/dt, f_1, g_1, f_2, g_2$ being rational functions of the differential operator D with constant coefficients and R_1, R_2 are functions of the independent variable t. We seek a solution of above pair of differential equations satisfying both equations. Operating these equations by $g_2(D)$ and $g_1(D)$ respectively and taking their difference, the terms containing variable y get eliminated:

$$g_2(D)\{f_1(D)\,x(t)\} - g_1(D)\{f_2(D)\,x(t)\}$$

$$\equiv \{g_2(D) f_1(D) - g_1(D) f_2(D)\}x(t)\} = g_2(D)\,R_1(t) - g_1(D)\,R_2(t), \quad (8.2)$$

where we have used the commutative property of the differential operators:

$$g_1(D)\,g_2(D)\,y(t) = g_2(D)\,g_1(D)\,y(t). \qquad (8.3)$$

The resulting equation (8.2) is a linear differential equation of some order depending upon the resultant power of D in the operator $(g_2 f_1 - g_1 f_2)$ with constant coefficients and can be solved by methods given in Section 4. Let the solution be $x = \psi(t)$. Substitution for this value of x in any of the two equations in (8.1) yields a differential equation in $y(t)$ and t:

$$g_1(D)\,y(t) = R_1(t) - f_1(D)\,\psi(t), \quad \text{or} \quad g_2(D)\,y(t) = R_2(t) - f_2(D)\,\psi(t),$$

which are also linear of the order depending on that of g_1, g_2 and can be solved by some method similarly.

8.2. Simultaneous differential equations in three variables: Let x, y and z be the rectangular Cartesian coordinates of a moving point in three-dimensional Euclidean space. The locus of such points is a surface represented by some functional relation

$$F(x, y, z) = 0. \qquad (8.4)$$

The number of independent variables involved in this equation is at most two and the remaining one as a function of the two independent variables. The first two coordinates: x and y are usually taken as the independent variables while z as a function of x and y. Let us consider the simultaneous equations

$$dx / P = dy / Q = dz / R, \qquad (8.5)$$

where P, Q, R are some functions of the variables x, y, z. The equations (8.5) determine the direction ratios of a tangent line to a curve. Taking any two ratios in Eq. (8.5) and integrating the resulting differential equations there result the surfaces represented by

$$u_1(x, y, z) = c_1 \quad \text{and} \quad u_2(x, y, z) = c_2\,, \qquad (8.6)$$

where c_1, c_2 are arbitrary constants of integration. The equations (8.6) holding simultaneously determine the curve of intersection of the (solution) surfaces. Since the arbitrary constants may be chosen in an infinite number of ways there exist doubly infinite number of such curves. There are different methods of integration of the equations (8.5) illustrated in the following examples.

Example 8.1. Solve the simultaneous differential equations

$$dx / z = dy / 0 = -dz / x. \tag{8.7}$$

Solution. The first and the third ratios determine the differential equation (in separated form of variables) $x.\ dx + z.\ dz = 0$ having integral

$$x^2 + z^2 = a^2, \tag{8.8}$$

where a is an arbitrary constant of integration. This equation represents a family of coaxial right circular cylinders. On the other hand, the second ratio with any other ratio in Eq. (8.7) constitutes the differential equation $dy = 0$ having solutions

$$y = b, \tag{8.9}$$

b being an arbitrary constant of integration. The equations (8.9) represent a family of planes in the space. Thus, the solutions given by Eqs. (8.8) and (8.9) holding simultaneously and satisfying the differential equations (8.7) represent a family of circles (obtained by intersection of the cylinders and the planes). //

Example 8.2. Solve the simultaneous differential equations

$$dx / yz = dy / zx = dz / xy. \tag{8.10}$$

Solution. Dropping the common multiple $1/z$ from the first two ratios there results the differential equation (in separated form of variables) $x.\ dx - y.\ dy = 0$ having integral

$$x^2 - y^2 = a^2, \tag{8.11}$$

where a is an arbitrary constant of integration. This equation represents a family of rectangular hyperbolic cylinders. Again, dropping the common multiple $1/y$ from the first and the last ratios there results the differential equation (in separated form of variables) $x.\ dx - z.\ dz = 0$ having integral

$$x^2 - z^2 = b^2, \tag{8.12}$$

b being an arbitrary constant of integration. This equation also represents another family of rectangular hyperbolic cylinders. These surfaces intersect in a family of curves representing the solution of Eq. (8.10). //

Example 8.3. Solve the simultaneous differential equations

$$dx / y = dy / x = dz / z. \tag{8.13}$$

Solution. The first two ratios form the differential equation $x.\, dx - y.dy = 0$, where the variables are separated and its integral is found vide equation (8.11). Considering the last two ratios and putting for x from Eq. (8.11), there results the differential equation (in separated variables form)

$$dy / \sqrt{(y^2 + a^2)} = dz / z\,,$$

having integral

$$\ln \{y + \sqrt{(y^2 + a^2)}\} = \ln z + \ln b = \ln (bz) \Rightarrow y + \sqrt{(y^2 + a^2)}\} = bz.$$

Eliminating the constant a from the two solutions, there results the common solution of the simultaneous equations (8.13): $x + y = bz$, representing a family of planes. //

Example 8.4. Solve the simultaneous differential equations

$$dx / z.\,(x + y) = dy / z.\,(x - y) = dz / (x^2 + y^2). \tag{8.14}$$

Solution. Using $x, - y, - z$ as the multipliers both in the numerator and denominator of respective fractions each of the fraction is also equal to

$$(x.\, dx - y.\, dy - z.\, dz) / \{x\, z.\, (x + y) - y\, z.\, (x - y) - z.\, (x^2 + y^2)\}$$

$$= (x.\, dx - y.\, dy - z.\, dz) / 0.$$

Thus, equating any fraction in Eq. (8.14) with this new fraction there results the differential equation $x.\, dx - y.\, dy - z.\, dz = 0$ having integral

$$x^2 - y^2 - z^2 = a^2, \tag{8.15}$$

a being an arbitrary constant of integration. This equation represents a family of rectangular hyperboloids of two sheets.

On the other hand, using y, x, $-z$ as the multipliers both in the numerator and denominator of respective fractions each of the fraction is also equal to

$$(y.\, dx + x.\, dy - z.\, dz) \,/\, \{y\, z.\, (x + y) + x\, z.\, (x - y) - z.\, (x^2 + y^2)\}$$

$$= (y.\, dx + x.\, dy - z.\, dz) \,/\, 0$$

$\Rightarrow$

$$y.\, dx + x.\, dy - z.\, dz = 0, \quad \text{i.e.} \quad 2\, d\,(x\, y) - 2z.\, dz = 0,$$

having integral

$$2\, x\, y - z^2 = b^2. \tag{8.16}$$

This equation represents a family of parabolic cylinders. Thus, the solution of Eq. (8.14) is a family of curves of intersection of the surfaces represented by Eqs. (8.15) and (8.16). //

Example 8.5. Solve the simultaneous differential equations

$$dx \,/\, (y^2 + yz + z^2) = dy \,/\, (z^2 + zx + x^2) = dz \,/\, (x^2 + xy + y^2). \tag{8.17}$$

Solution. Each of the ratio is also equal to

$$(dx - dy) \,/\, \{(y^2 - x^2) + z\,(y - x)\} = (dy - dz) \,/\, \{(z^2 - y^2) + x\,(z - y)\}$$

$$= (dz - dx) \,/\, \{(x^2 - z^2) + y\,(x - z)\},$$

or,

$$(dx - dy) \,/\, \{(y - x)\,(y + x + z)\} = (dy - dz) \,/\, \{(z - y)\,(z + y + x)\}$$

$$= (dz - dx) \,/\, \{(x - z)\,(x + z + y)\},$$

or, dropping the common factor $-(x + y + z)$ from the denominator in each ratio

$$(dx - dy) \,/\, (x - y) = (dy - dz) \,/\, (y - z) = (dz - dx) \,/\, (z - x). \tag{8.18}$$

Taking the first two ratios and integrating the differential equation so formed, we get

$$\ln\,(x - y) = \ln\,(y - z) + \ln\, a \equiv \ln\, \{a\,(y - z)\}$$

$\Rightarrow$

$$x - y = a\,(y - z), \tag{8.19}$$

a being an arbitrary constant of integration.

On the other hand, taking the last two ratios in Eq. (8.18) and integrating the differential equation so formed, we get

$$\ln (y-z) = \ln (z-x) + \ln b \equiv \ln \{b(z-x)\} \Rightarrow y-z = b(z-x), \quad (8.20)$$

b being an arbitrary constant of integration. The solution of (8.17) is a family of curves of intersection of the surfaces represented by Eqs. (8.19) and (8.20). //

Example 8.6. Solve the simultaneous differential equations

$$x.\,dx / (z^2 - 2yz - y^2) = dy / (y+z) = dz / (y-z). \quad (8.21)$$

Solution. Using 1, y and z as the multipliers both in the numerator and denominator of respective fractions each of the fraction is also equal to

$$(x.\,dx + y.\,dy + z.\,dz) / \{(z^2 - 2yz - y^2) + y.\,(y+z) + z.\,(y-z)\}$$

$$= (x.\,dx + y.\,dy + z.\,dz) / 0.$$

Thus, equating any fraction in Eq. (8.21) with this new fraction there results the differential equation $x.\,dx + y.\,dy + z.\,dz = 0$ having integral

$$x^2 + y^2 + z^2 = a^2, \quad (8.22)$$

a being an arbitrary constant of integration. This equation represents a family of spheres.

On the other hand, the last two ratios in Eq. (8.21) form a differential equation

$$(y-z).\,dy = (y+z).\,dz, \text{ or } y.\,dy - z.\,dz = y.\,dz + z.\,dy$$

or, equivalently

$$d(y^2 - z^2) = 2\,d(y\,z),$$

having integral

$$y^2 - z^2 = 2\,y\,z + b, \quad (8.23)$$

b being an arbitrary constant of integration. The solution of (8.21) is a family of curves of intersection of the surfaces represented by Eqs. (8.22) and (8.23). //

Example 8.7. Find the curve through the point (1, 2, – 1) whose tangent at a point (x, y, z) has direction ratios which are the squares of the coordinates of the point.

Solution. Since dx, dy and dz form the direction ratios of the tangent to the curve at the point (x, y, z); which, as per hypothesis, are proportional to x^2, y^2, z^2 there exist the simultaneous differential equations

$$dx / x^2 = dy / y^2 = dz / z^2. \tag{8.24}$$

Taking the first two ratios and integrating the differential equation so obtained, we get $1/x - 1/y = a$, whereas the differential equation formed by the last two ratios has integral $1/y - 1/z = b$. The curve of intersection of these surfaces represents the solution of Eq. (8.24). By hypothesis, the surfaces pass through the given point (1, 2, – 1). Hence, we have

$$1 - 1/2 = 1/2 = a \quad \text{and} \quad 1/2 + 1 = 3/2 = b.$$

Thus, the solution curve is

$$1/x - 1/y = 1/2 \qquad \text{together with} \qquad 1/y - 1/z = 3/2,$$

or, equivalently

$$1/x = 1/y + 1/2 = 1/z + 2. \text{ //}$$

§ 9. Solution of ODE by Variation of Parameters Method

9.1. Solution of a linear ODE of the first order

Consider a differential equation

$$dy / dx + P(x).\, y = Q(x). \tag{9.1}$$

Its associated *homogenous* form

$$dy / dx + P(x).\, y = 0, \quad \text{or} \quad dy / y + P(x)\, dx = 0,$$

has integral

$$\ln y + \int P(x)\, dx = \ln c, \quad \text{or} \quad y. \exp\{\int P(x)\, dx\} = c$$

⇒

$$y_c = c. \exp\{-\int P(x)\, dx\}, \tag{9.2}$$

where c is an arbitrary constant of integration. Presuming that replacement of the constant c by some suitable function $u(x)$ in (9.2) provides a solution of the original differential equation (9.1):

$$y_P = u(x).\ \exp\{-\int P(x)\, dx\}, \tag{9.3}$$

so as to the solution y_P with its derivative

$$dy_P / dx = u'(x).\ \exp\{-\int P(x)\, dx\} - u.\ P.\ \exp\{-\int P(x)\, dx\} \tag{9.4}$$

satisfy the original diff. Eq. (9.1). Thus, putting from (9.3) and (9.4), the Eq. (9.1) turns

$$u'(x).\ \exp\{-\int P(x)\, dx\} = Q(x) \quad \Rightarrow \quad u'(x) = Q(x).\ \exp\{\int P(x)\, dx\}.$$

Integrating it, we thus evaluate the function $u(x)$:

$$u(x) = \int [Q(x).\ \exp\{\int P(x)\, dx\}].\ dx. \tag{9.5}$$

The sum of functions y_c and y_P :

$$y = y_c + y_P = [c + \int Q(x).\ \exp\{\int P(x)\, dx\}].\ \exp\{-\int P(x)\, dx\},$$

i.e.

$$y.\ \exp\{\int P(x)\, dx\} = \int Q(x).\ \exp\{\int P(x)\, dx\} + c \tag{9.6}$$

provides the complete soln. of Eq. (9.1).

Note 9.1. The method (demonstrated above) is not preferable for the solution of 1st order linear ODE as there is a much shorter method by introduction of an integrating factor

$$\text{I.F.} = \exp\{\int P(x)\, dx\}.$$

Multiplying the ODE (9.1) by this I.F.:

$$\{dy/dx + P(x).\ y\}.\ \exp\{\int P(x)\, dx\} \equiv (d/dx)\ \{y.\ \exp\{\int P(x)\, dx\}$$

$$= Q(x).\ \exp\{\int P(x)\, dx\},$$

and integrating the resulting equation, one immediately gets the solution given by Eq. (9.6).

9.2. Solution of linear ODE of 2nd order with constant coefficients

Consider an ODE

$$a_0\,(d^2y\,/\,dx^2) + a_1\,(dy\,/\,dx) + a_2 . y \;=\; Q_1\,(x).$$

Dividing it by $a_0 \neq 0$, and putting $P \equiv a_1/a_0$, $Q \equiv a_2/a_0$, $R \equiv Q_1/a_0$, it may be re-written as

$$y'' + P . y' + Q . y \;=\; R\,(x) \tag{9.7}$$

where primes denote derivation w.r.t. x. Its associated homogenous eqn. (with vanishing RHS)

$$y'' + P . y' + Q . y = 0 \tag{9.8}$$

gives rise to its auxiliary equation $m^2 + Pm + Q = 0$ for some (algebraic) number m. Its roots m_1, m_2 determine the complimentary function C.F. [cf. § 4]:

$$y_c \;=\; A_1 . \exp\,(m_1 x) + A_2 . \exp\,(m_2 x), \tag{9.9}$$

where A_1, A_2 are arbitrary constants of integration. It is also presumed that the roots m_1, m_2 are distinct and real (numbers). In general, let the ODE (9.8) possesses two distinct solutions, say y_1 and y_2, so that above C.F. assumes the form

$$y_c \;=\; A_1 . y_1 + A_2 . y_2. \tag{9.10}$$

Like Case 9.1, let replacement of the constants A_1, A_2 by some functions $u_1\,(x)$ and $u_2\,(x)$ in (9.10) provides a solution of the original differential equation (9.7):

$$y_P \;=\; u_1 . y_1 + u_2 . y_2\,, \tag{9.11}$$

so that the soln. y_P and its derivatives:

$$y'_P = u'_1 . y_1 + u_1 . y_1' + u'_2 . y_2 + u_2 . y_2',$$

$$y''_P = (u''_1 . y_1 + 2u'_1 . y_1' + u_1 . y_1'') + (u''_2 . y_2 + 2u'_2 . y_2' + u_2 . y_2'')$$

should satisfy the original diff. Eq. (9.7). Thus, putting these values, the Eq. (9.7) yields

$$u_1 .\{y_1'' + P . y_1' + Q . y_1\} + u_2 .\{y_2'' + P . y_2' + Q . y_2\} + (u''_1 . y_1 + u'_1 . y_1' +$$

$$u''_2 . y_2 + u'_2 . y_2') + P\,(u'_1 . y_1 + u'_2 . y_2) + (u'_1 . y_1' + u'_2 . y'_2) \;=\; R\,(x).$$

As per hypothesis, both y_1, y_2 being solutions of homogeneous ODE (9.8), make the coefficients of u_1, u_2 in above equation zero reducing above equation:

$$(d/dx)\,(u'_1 . y_1 + u'_2 . y_2) + P\,(u'_1 . y_1 + u'_2 . y_2) + (u'_1 . y_1' + u'_2 . y'_2) = R\,(x).$$

Choosing the functions u_1, u_2 so as to satisfy

$$u'_1 . y_1 + u'_2 . y_2 = 0, \tag{9.12}$$

above equation further simplifies to

$$u'_1 . y_1' + u'_2 . y'_2 = R\,(x). \tag{9.13}$$

Solving these simultaneous linear eqns. for u'_1, u'_2:

$$u'_1 \Big/ \begin{vmatrix} y_2 & 0 \\ y_2' & -R \end{vmatrix} - u'_2 \Big/ \begin{vmatrix} y_1 & 0 \\ y_1' & -R \end{vmatrix} = 1 \Big/ \begin{vmatrix} y_1 & y_2 \\ y_1' & y_2' \end{vmatrix}$$

$\Rightarrow$

$$u'_1 = -R\,y_2 / (y_1 y'_2 - y'_1 y_2) \quad \text{and} \quad u'_2 = R\,y_1 / (y_1 y'_2 - y'_1 y_2),$$

and integrating the last two equations, we evaluate the functions u_1, u_2. Thus, the general soln. of the original ODE (9.7) is obtained:

$$y = y_c + y_P = (A_1 y_1 + A_2 y_2) + (u_1 y_1 + u_2 y_2). \tag{9.14}$$

Example 9.1. Solve the ODE

$$y'' - 4y' + 4y = (x + 1)\,e^{2x}, \tag{9.15}$$

by using the method of variation of parameters.

Solution. Auxiliary equation for given ODE is $m^2 - 4m + 4 = 0$, having a double (repeated) root $m = 2$. Hence, the C.F. for the solution of associated homogeneous form of (9.15) is

$$y_c = (A_1 + A_2\,x)\,e^{2x},$$

and the particular integral is

$$y_P = (u_1 + u_2\,x)\,e^{2x},$$

where u_1, u_2 are some suitable functions so that y_P and its derivatives:

$$y'_P = (u'_1 + u'_2 . x + u_2)\, e^{2x} + 2\,(u_1 + u_2\, x)\, e^{2x},$$

$$y''_P = (u''_1 + 2u'_2 + u''_2 . x)\, e^{2x} + 4(u'_1 + u'_2 . x + u_2)\, e^{2x} + 4\,(u_1 + u_2\, x)\, e^{2x}$$

should satisfy the original diff. Eq. (9.15). Thus, putting these values, the Eq. (9.15) yields

$$u''_1 + u''_2 . x + 2u'_2 = x + 1, \quad \text{or} \quad (d/dx)\,(u'_1 + u'_2 . x) + u'_2 = x + 1.$$

Choosing the functions u_1, u_2 so as to satisfy

$$u'_1 + u'_2 . x = 0, \tag{9.16}$$

above equation reduces to

$$u'_2 = x + 1. \tag{9.17}$$

Putting from (9.17) in (9.16), we get

$$u'_1 = -\, x\,(x + 1) = -x^2 - x. \tag{9.18}$$

Integrating the last two equations, we evaluate the functions u_1, u_2:

$$u_1 = -x^3/3 - x^2/2 \qquad \text{and} \qquad u_2 = x^2/2 + x.$$

Hence , $y_P = (-x^3/3 - x^2/2 + x^3/2 + x^2)\, e^{2x} = (x^3/6 + x^2/2)\, e^{2x}$.

And, therefore, the general soln. of the ODE (9.15) is

$$y = y_c + y_P = (A_1 + A_2 . x + x^2/2 + x^3/6)\, e^{2x}. \;//$$

§ 10. Problem Set

10.1. Solve the following differential equations of first order and first degree:

(i) $x^2\,(1 - y)\, dy + y^2\,(1 + x)\, dx = 0$, **(ii)** $dy/dx = e^{x-y} + x^2 e^{-y}$,

(iii) $(x . y^2 + x)\, dx + (x^2 . y + y)\, dy = 0$, **(iv)** $(1 - x^2)\, dy/dx + x\, y = x\, y^2$.

[**Hint:** Division by $x^2 y^2$, e^{-y}, $(1 + x^2)(1 + y^2)$ and $(1 - x^2)\, y\, (1 - y)$ in respective equations turns them into separated variables forms.]

10.2. Solve the following (homogeneous) differential equations:

(i) $dy/dx = (x^2 + y^2)/2x^2$, **(ii)** $(x^2 - y^2)\, dy = x\, y.\, dx$,

(iii) $dy/dx = y/x + \tan(y/x)$, **(iv)** $(x^2.y)\, dx - (x^3 + y^3)\, dy = 0$,

(v) $x.\cos(y/x)\,(y.dx + x.dy) = y.\sin(y/x)\,(x.dy - y.dx)$.

10.3. Solve the following differential equations which are reducible to homogeneous forms:

(i) $dy/dx = (y - x + 1)/(y + x + 5)$, **(ii)** $(x + 2y)(dx - dy) = dx + dy$,

(iii) $(x - y - 2)\, dx + (x - 2y - 3)\, dy = 0$,

(iv) $2(x - 2y + 1)\, dy/dx = 4x - 2y + 1$.

[**Hint: (ii)** Usual method of Sub-section 2.3 fails here. Putting $x + 2y = v$ reduces the equation to

$$dv/dx = (3v - 1)/(v + 1), \quad \text{or} \quad \{(v + 1)/(3v - 1)\}\, dv = dx,$$

where the variables are separated. Breaking

$$(v + 1)/(3v - 1) \quad \text{as} \quad (1/3)\{1 + 4/(3v - 1)\},$$

an integral of above equation is

$$3v + 4\ln(3v - 1) = 9x + 2a, \quad \text{or} \quad 2\ln(3x + 6y - 1) = 3(x - y) + a\,].$$

10.4. Solve the following linear differential equations:

(i) $dy/dx + 2x\, y = 2e^{-x^2}$, **(ii)** $dy/dx + 2y/x = \sin x$

(iii) $x.(1 - x^2)\, dy + (2x^2.\, y - y - ax^3)\, dx = 0$,

(iv) $(1 + y^2)\, dx = (\tan^{-1} y - x)\, dy$,

(v) $y + d(xy)/dx = x.(\sin x + \ln x)$, **(vi)** $y - x\,(dy/dx) = x + dy/dx$.

10.5. Solve the following differential equations which are reducible to a linear form:

(i) $x.\, dy/dx + y = y^2.\ln x$ **(ii)** $(x^3.y^3 + xy)\, dy - dx = 0$

(iii) $2\,(dy/dx) - y.\sec x = y^3.\tan x$,

(iv) $dy/dx - (\tan y)/(1 + x) = (1 + x)\, e^x.\sec y$,

(v) $\cos x.\, dy = y.(\sin x - y).\, dx$,

(vi) $(\sin x)\, dy/dx + y.\cos x = 2\sin^2 x.\cos x$.

10.6. Solve the following exact differential equations:

(i) $(x^3 + y^3)\, dx + (3\, xy^2)\, dy = 0$, **(ii)** $(y.\sin 2x)\, dx - (y^2 + \cos^2 x)\, dy = 0$,

(iii) $dy/dx = (2x - y + 1)/(x + 2y - 3)$,

(iv) $(2xy + y - \tan y)\, dx + (x^2 - x.\tan^2 y + \sec^2 y)\, dy = 0$,

(v) $\{y\,(1 + x^{-1}) + \cos y\}dx + \{x + \ln x - x.\sin y\}\, dy = 0$.

10.7. Find the suitable integrating factor and solve the differential equation

$$\{x\,y.\sin(x\,y) + \cos(x\,y)\}y.\, dx + \{x\,y.\sin(x\,y) - \cos(x\,y)\}x.\, dy = 0.$$

[**Hint:** Similar to Example 2.3,

$$\text{I. F.} = 1/(f_1 - f_2)\,x\,y = 1/\,2x\,y.\cos(x\,y).$$

Multiplication by the same makes the equation in exact form:

$$\{\tan(x\,y) + 1/x\,y\}\; y.\, dx + \{\tan(x\,y) - 1/\,xy\}x.\, dy = 0,$$

having solution

$$\ln\sec(x\,y) + \ln x - \ln y = \ln a, \qquad \text{or} \qquad x.\sec(x\,y) = a\,y\;].$$

10.8. Solve the following differential equations by a suitable change of variables:

(i) $\cos(x + y)\, dy = dx$,

(ii) $(x\, dx + y\, dy) / (x\, dy - y\, dx) = \sqrt{(a^2 - x^2 - y^2)} / \sqrt{(x^2 + y^2)}$,

(iii) $x\,(dy/dx) - y + \sqrt{(y^2 + 2x^2)} = 0$, **(iv)** $(x + y - 1)\, dy = (x + y)\, dx$.

[Hint: (ii) Changing x, y to polar coordinates $x = r\cos\theta$, $y = r\sin\theta$ so that

$$x^2 + y^2 = r^2, \qquad \text{and} \qquad y / x = \tan\theta.$$

Their derivatives yield

$$x\, dx + y\, dy = r\, dr, \qquad \text{and} \qquad (x\, dy - y\, dx) / x^2 = \sec^2\theta.\, d\theta$$

$\Rightarrow$

$$x\, dy - y\, dx = r^2.\, d\theta,$$

which transform the equation to $dr / d\theta = \sqrt{(a^2 - r^2)}$. Separating the variables and integrating the resulting equation w.r.t. respective variables, there follows

$$\sin^{-1}(r / a) = \theta + b \qquad \Rightarrow \qquad r = a\sin(\theta + b).$$

Changing back to Cartesian coordinates, above relation assumes the form

$$\sqrt{(x^2 + y^2)} = a\sin\{\tan^{-1}(y / x) + b\}.$$

(iii) Rewriting the equation as $dy/dx = \{y - \sqrt{(y^2 + 2x^2)}\}/x$ and following the procedure of Sub-section 2.2, it transforms to a separated variables form:

$$dv / \sqrt{(v^2 + 2)} + dx / x = 0.$$

Further, putting

$$v = \sqrt{2}.\tan\theta, \qquad \text{so that} \qquad dv = \sqrt{2}.\sec^2\theta.\, d\theta,$$

above equation reduces to $\sec\theta.\, d\theta + dx / x = 0$, having integral

$$\ln(\sec\theta + \tan\theta) + \ln x = \ln a \qquad \Rightarrow \qquad x(\tan\theta + \sec\theta) = a.$$

Reverting θ back to *v* and subsequently to *y*, the integral equation reduces to

$$x\,\{v/\sqrt{2} + \sqrt{(1 + v^2/2)}\} = y/\sqrt{2} + \sqrt{(x^2 + y^2/2)} = a$$

$\Rightarrow$

$$(y - a\sqrt{2})^2 = 2x^2 + y^2.\,]$$

10.9. Solve the differential equation $dy/dx = x/(x^2 y + y^3)$ when $y(0) = 0$.

[Hint: Rewriting the equation as $x\,(dx/dy) - x^2 y = y^3$, and putting $x^2 = t$ so that $2x\,(dx/dy) = dt/dy$, the equation becomes linear in *t* and *dt*/*dy*:

$$dt/dy - (2y)\,t = 2y^3$$

with an integrating factor $e^{-\int 2y\,dy} = e^{-y^2}$, and solution

$$t.\,e^{-y^2} = 2\int y^3.\,e^{-y^2}.\,dy + a. \qquad (10.1)$$

Putting $y^2 = u$ so that $2y\,dy = du$, the integral in the RHS reduces to $\int u.e^{-u}.\,du$. Applying method of integration by parts taking e^{-u} as the second function the integral evaluates as:

$$-u.\,e^{-u} + \int e^{-u}.\,du = -(u+1)\,e^{-u} = -(y^2+1)\,e^{-y^2}.$$

Hence, Eq. (10.1) when multiplied by e^{y^2} yields the solution $x^2 + y^2 + 1 = a.e^{y^2}$. For the initial conditions $x = 0$, $y = 0$, the constant of integration *a* becomes 1 and the solution simplifies to $x^2 + y^2 + 1 = e^{y^2}$.]

10.10. If $dy/dx = e^{-2y}$ and $y = 0$ when $x = 5$, then find the value of x at $y = 3$.

10.11. Solve the following differential equations, where $p = dy/dx$:

(i) $p^2 + px + py + xy = 0$, **(ii)** $xy\,(y - px) = x + py$, **(iii)** $e^{p-y} = p^2 - 1$,

(iv) $xp^3 = a + bp$, **(v)** $y = p^2.\,y + 2px$, **(vi)** $y = px + \cos p$.

[**Hint: (i)** Factorizing the expression on LHS:

$$(p + x).\,(p + y) = 0 \quad \Rightarrow \quad p + x = 0, \qquad p + y = 0.$$

Separating the variables and integrating these differential equations, we have

$$\int dy + \int x\,dx = a, \quad \text{or,} \quad y + x^2/2 = a/2 \quad \Rightarrow \quad 2y + x^2 = a,$$

and

$$\int (1/y)\,dy + \int dx = a, \qquad \text{or,} \qquad \ln y + x = a.$$

Thus, the combined general solution of the differential equation is

$$(2y + x^2 - a).\,(\ln y + x - a) = 0.$$

(ii) Solution of the equation for p yields

$$p \equiv dy\,/\,dx = x\,(y^2 - 1)\,/\,(x^2 + 1)\,y.$$

Separating the variables:

$$\{y\,/\,(y^2 - 1)\}\,dy = \{x\,/\,(x^2 + 1)\}\,dx,$$

and term-wise integration gives

$$\ln (y^2 - 1) = \ln (x^2 + 1) + \ln a \quad \Rightarrow \quad y^2 = a\,(x^2 + 1) + 1.$$

(iii) Taking natural logarithm, the equation is solvable for y:

$$y = p - \ln (p^2 - 1). \tag{10.2}$$

Its derivation w.r.t. x yields:

$$p = \{1 - 2p\,/\,(p^2 - 1)\}\,(dp/dx).$$

Splitting $-2\,/\,(p^2 - 1)$ as $1\,/\,(p + 1) - 1\,/\,(p - 1)$ and separating the variables, we get:

$$\{1/\,p + 1/\,(p + 1) - 1/\,(p - 1)\}\,dp = dx.$$

Its term wise integral yields

$$x = \ln p + \ln (p + 1) - \ln (p - 1) + a,$$

which along with Eq. (10.2) constitute a general solution.

(iv) Solving the equation for x: $x = a/p^3 + b/p^2$, and differentiating it w.r.t. y, we get

$$dx/dy \equiv 1/p = -(3a/p^4 + 2b/p^3)\,(dp/dy), \text{ or } \quad dy = -(3a/p^3 + 2b/p^2)\,dp.$$

Integrating it term by term, there results a value for y in terms of p:

$$y = 3a/2p^2 + 2b/p + a.$$

The values of x and y expressed in terms of the parameter p constitute a general solution.

(v) Solving the equation for x : $x = y/2p - p\,y/2$ and differentiating it w.r.t. y, we get

$$dx/dy \equiv 1/p = 1/2p - p/2 - (y/2)\,(1/p^2 + 1)\,(dp/dy).$$

On multiplication by $2p^2$,

$$p\,(1 + p^2) + y\,(1 + p^2)\,(dp/dy) = 0,$$

division by $1 + p^2$ and separating the variables, there results the differential equation $dy/y + dp/p = 0$. Integrating it term-wise, there results

$$\ln y + \ln p = \ln a \quad \Rightarrow \quad p\,y = a \quad \Rightarrow \quad p = a/y.$$

Eliminating p from the given differential equation, we derive a general solution:

$$y^2 = a^2 + 2a\,x.$$

(vi) The equation is in Clairaut's form having a general solution $y = ax + \cos a$.]

10.12. Solve the following linear differential equations of order more than one with constant coefficients, where $D \equiv d/dx$:

(i) $(D^4 - 2D^3 + 2D^2 - 2D + 1)\,y = 0$, **(ii)** $(D - 3)^3\,(D + 3)^2\,y = 0$.

[**Hint: (i)** Auxiliary equation is $m^4 - 2m^3 + 2m^2 - 2m + 1 \equiv (m-1)^2.(m^2 + 1) = 0$, having two (repeated) real and two complex roots: $m = 1, 1, \pm i$. Hence, by Eqs. (4.6) and (4.8b), the desired solution is

$$y = (A_1 + A_2\,x)\,e^x + (A_3 \cos x + A_4 \sin x).$$

(ii) Auxiliary equation $(m-3)^3\,(m+3)^2 = 0$ has roots: $m = 3$ (repeating thrice), and -3 (repeating twice). Hence, by Eqs. (4.6) and (4.7), the desired solution is

$$y = (A_1 + A_2\,x + A_3\,x^2)\,e^{3x} + (A_4 + A_5\,x)\,e^{-3x}.]$$

10.13 Evaluating their "particular integral" solve the following linear differential equations of order more than one with constant coefficients, where $D \equiv d/dx$:

(i) $D^2\,(D+1)^2\,(D^2+D+1)^2\,y = e^x$, **(ii)** $(D^2+D-2)y = x + \sin x$,

(iii) $(D^2 + \omega_0)y = a.\cos\omega x$, when $\omega \to \omega_0$;

(iv) $(D^2 - 4D + 3)\,y = e^{3x}$, **(v)** $(D^3 - 3D - 2)y = 540.\,x^3.\,e^{-x}$,

(vi) $(D^4 + 2D^2 + 1)y = x^2.\cos x$, **(vii)** $(D^2 - 2D + 1)y = (x.\sin x)\,e^x$.

[**Hint: (i)** Auxiliary equation $m^2\,(m+1)^2\,(m^2+m+1)^2 = 0$ has two real and two complex roots (each repeating twice): $m = 0, -1, (-1 \pm i\sqrt{3})/2$. Hence, by Eqs. (4.6) and (4.9), we have

$$\text{C.F.} = (A_1 + A_2\,x) + (A_3 + A_4\,x)\,e^{-x} +$$

$$\{\, (A_5 + A_6\,x)\cos(x\sqrt{3}/2) + (A_7 + A_8\,x)\sin(x\sqrt{3}/2)\,\}\,e^{-x/2}.$$

Also, by Eq. (5.6),

$$\text{P.I.} = \{1/D^2\,(D+1)^2\,(D^2+D+1)^2\}\,e^x = e^x/\,1^2.2^2.3^2 = e^x/36$$

determining the solution $y = \text{C.F.} + \text{P.I.}$

(ii) The auxiliary equation $m^2 + m - 2 = 0$ has distinct roots $m = 1, -2$. Hence, by Eq. (4.3), we have

$$\text{C.F.} = A_1\,e^{x} + A_2\,e^{-2x}.$$

The particular integral may be evaluated in view of Note 5.1 and Eq. (5.8):

$$1/(D^2+D-2) = -(1/2)\,\{1 - (D+D^2)/2\}^{-1}$$

$$= -(1/2)\{1 + (D + D^2)/2 + \ldots$$

$\Rightarrow$

$$\{1/(D^2 + D - 2)\}\, x = -(1/2)\,(x + 1/2),$$

and

$$\{1/(D^2 + D - 2)\} \sin x = \{1/(D - 3)\} \sin x$$

$$= e^{3x}. \int e^{-3x}. \sin x.\, dx, \qquad \text{by Eq. (5.4)}$$

$$= -(3.\sin x + \cos x)/(9 + 1) = -(3.\sin x + \cos x)/10, \text{ by Eq. (5.17).}$$

Hence, the solution is

$$y = \text{C.F.} + \text{P.I.} = A_1 e^{x} + A_2 e^{-2x} - (2x + 1)/4 - (3.\sin x + \cos x)/10.$$

(iii) The auxiliary equation $m^2 + \omega_0 = 0$ has imaginary roots $m = \pm i \sqrt{\omega_0}$. Hence, by Eq. (4.8b), we have

$$\text{C.F.} = A_1 \cos (x \sqrt{\omega_0}) + A_2 \sin (x \sqrt{\omega_0}).$$

Also, by Eq. (5.8),

$$\text{P.I.} = \{1/(D^2 + \omega_0)\}\,(a. \cos \omega x) = (a. \cos \omega x)/(\omega_0 - \omega^2)$$

$$= (a. \cos \omega_0 x)/\omega_0 (1 - \omega_0),$$

when $\omega \rightarrow \omega_0$. Hence, the solution is

$$y = \text{C.F.} + \text{P.I.} = A_1 \cos (x\sqrt{\omega_0}) + A_2 \sin (x\sqrt{\omega_0}) + (a. \cos \omega_0 x)/\omega_0 (1 - \omega_0).$$

(iv) The auxiliary equation $m^2 - 4m + 3 \equiv (m - 1)(m - 3) = 0$ has distinct real roots $m = 1, 3$. Hence, by Eq. (4.3), we have

$$\text{C.F.} = A_1 e^{x} + A_2 e^{3x}.$$

The particular integral $1/(D^2 - 4D + 3)\}\, e^{3x}$ cannot be evaluated by Eq. (5.6) as $D^2 - 4D + 3$ vanishes at $x = 3$. So, in view of Note 5.3, it is evaluated as per Theo. 5.5, by taking $u(x) = 1$:

$$\text{P.I.} = e^{3x}. (D + 3 - 1)^{-1} (D + 3 - 3)^{-1} (1) = (e^{3x}/2).(1 + D/2)^{-1} (x)$$

$$= (e^{3x}/2).\,(1 - D/2 + D^2/4 - \ldots)\,(x) = (e^{3x}/2).\,(x - 1/2).$$

The last term in P.I. being already included in the C.F. is dropped and the solution is

$$y = \text{C.F.} + \text{P.I.} = A_1\, e^x + (A_2 + x/2)\, e^{3x}.$$

(v) The auxiliary equation $m^3 - 3m - 2 \equiv (m+1)^2\,(m-2) = 0$ has one repeated root $m = -1$, and $m = 2$. Hence, by Eqs. (4.3) and (4.6), we have

$$\text{C.F.} = (A_1 + A_2 x)\; e^{-x} + A_3\, e^{2x}.$$

Also, by Eq. (5.14),

$$\text{P.I.} = \{1/(D+1)^2\,(D-2)\}(540.x^3.e^{-x}) = 540.\; e^{-x}\,\{1/D^2\,(D-3)\}(x^3)$$

$$= -\,180.\; e^{-x}.(1\,/\,D^2)\,(1 - D/3)^{-1}\,(x^3)$$

$$= -\,180.\; e^{-x}.\,(1/D^2\,)\,(1 + D/3 + D^2/9 + D^3/27 + \dots)\,(x^3)$$

$$= -\,180.e^{-x}.(1/D^2 + 1/3D + 1/9 + D/27 + D^2/81 + D^3/243 + \dots)\,(x^3)$$

$$= -\,180.e^{-x}.(x^5/20 + x^4/12 + x^3/9 + x^2/9 + 2x/27 + 2/81).$$

The last two terms in P.I. being already included in the C.F. are dropped and the solution is

$$y = \text{C.F.} + \text{P.I.} = \{A_1 + A_2 x - (20.x^2 + 20x^3 + 15x^4 + 9x^5)\}\; e^{-x} + A_3\, e^{2x}.$$

(vi) The auxiliary equation $m^4 + 2m^2 + 1 \equiv (m^2+1)^2 = 0$ has repeated imaginary roots $m = \pm\, i$. Hence, by Eq. (4.9), we have

$$\text{C.F.} = (A_1 + A_2 x)\cos x + (A_3 + A_4 x)\sin x.$$

Also, by Eq. (5.14),

$$\text{P.I.} = \{1/\,(D^2+1)^2\}\;(x^2.\cos x) = \{1\,/\,(D^2+1)^2\}\,\{\text{Rl}\;\;(x^2.\,e^{\,ix})\}$$

$$= \text{Rl}\,[\,e^{\,ix}.\;\{(D+i\,)^2+1\}^{-2}\,(x^2)\,] \;=\; \text{Rl}\;\{e^{\,ix}.\,(D^2+2iD)^{-2}\,(x^2)\}$$

$$= \text{Rl}\;\{e^{\,ix}.\,(-1/4D^2)\,(1 - iD\,/2)^{-2}\,(x^2)\}$$

$$= \;\text{Rl}\;\{e^{\,ix}.(-1/4D^2)\,(1 + i\,D - 3D^2/4 - i\,D^3/2 + 5D^4/16 + \dots)\,(x^2)$$

$$= \;\text{Rl}\;\{\,e^{\,ix}.(-\,1/4D^2 - i\,/4D + 3/16 + i\,D/8 - 5D^2/64 + \dots)\,(x^2)$$

$$= \text{Rl}\ \{(\cos x + i \sin x)(-x^4/48 - ix^3/12 + 3x^2/16 + i\,x/4 - 5/32)\}$$

$$= (-x^4/48 + 3x^2/16 - 5/32)\cos x + (x^3/12 - x/4)\sin x.$$

The terms in P.I. being already included in the C.F. are dropped and the solution is

$$y = \text{C.F.} + \text{P.I.}$$

$$= (A_1 + A_2 x + 3x^2/16 - x^4/48)\cos x + (A_3 + A_4 x + x^3/12)\sin x.$$

(vii) The auxiliary equation $m^2 - 2m + 1 \equiv (m-1)^2 = 0$ has a double (real) root $m = 1$. Hence, by Eq. (4.6), we have

$$\text{C.F.} = (A_1 + A_2 x)\, e^x.$$

Also, by Eq. (5.14),

$$\text{P.I.} = \{1/(D-1)^2\}(xe^x \sin x) = e^x(1/D^2)(x.\sin x) = e^x(1/D^2)\{\text{Im}(x\, e^{ix})\}$$

$$= e^x.\ \text{Im}\ [e^{ix}\{1/(D+i)^2\}(x)] = -e^x.\ \text{Im}\ \{e^{ix}(1 - iD)^{-2}(x)\}$$

$$= -e^x.\ \text{Im}\ \{e^{ix}.(1 + 2iD - \ldots)(x)\}$$

$$= -e^x.\ \text{Im}\ \{(\cos x + i\sin x)(x + 2i)\} = -e^x.(2\cos x + x.\sin x).$$

Therefore, the solution is

$$y = \text{C.F.} + \text{P.I.} = \{A_1 + A_2 x - (2\cos x + x.\sin x)\}.\ e^x.\ ]$$

10.14. Solve the following homogeneous linear differential equations with variable coefficients, where $D \equiv d/dx$:

(i) $(x^2.D^2 + x.D + 1)\,y = x$, **(ii)** $(x^2 D^2 + 5x D + 4)\,y = x^4$,

(iii) $(x^2 D^2 - x D + 1)\,y = x^2$, when $y = 1$ and $dy/dx = 0$ at $x = 1$,

(iv) $\{(x+a)^2.D^2 - 4(x+a).D + 6\}\,y = x$.

[**Hint: (i)** Under the change of parameters given by Eq. (6.2), the differential equation reduces to

$$\{D'(D'-1) + D' + 1\}\,y \equiv (D'^2 + 1)\,y = e^t.$$

Its auxiliary equation $m^2 + 1 = 0$ has imaginary roots $m = \pm i$. Hence, by Eq. (4.8b), we get

$$\text{C.F.} = A_1 \cos t + A_2 \sin t = A_1 \cos(\ln x) + A_2 \sin(\ln x).$$

Also, by Eq. (5.6),

$$\text{P.I.} = \{1/(D'^2 + 1)\}(e^t) = e^t/2 = x/2.$$

Hence, the solution is

$$y = \text{C.F.} + \text{P.I.} = A_1 \cos(\ln x) + A_2 \sin(\ln x) + x/2.$$

(ii) As above, under the change of parameters given by Eq. (6.2), the differential equation transforms to

$$\{D'(D' - 1) + 5D' + 4\}\, y \equiv (D'^2 + 4D' + 4)\, y = e^{4t}.$$

Its auxiliary equation $m^2 + 4m + 4 \equiv (m + 2)^2 = 0$ has a double root $m = -2$. Hence, by Eq. (4.6), we get

$$\text{C.F.} = (A_1 + A_2 t)\, e^{-2t} = (A_1 + A_2 \ln x)\, x^{-2}.$$

Also, by Eq. (5.6),

$$\text{P.I.} = \{1/(D' + 2)^2\}\, e^{4t} = e^{4t}/(4 + 2)^2 = e^{4t}/36 = x^4/36.$$

Hence, the solution is

$$y = \text{C.F.} + \text{P.I.} = (A_1 + A_2 \ln x)\, x^{-2} + x^4/36.$$

(iii) As the above case, the equation reduces to the form

$$\{D'(D' - 1) - D' + 1\}\, y \equiv (D'^2 - 2D' + 1)\, y = e^{2t}.$$

Its auxiliary equation $m^2 - 2m + 1 = 0$ has a double root $m = 1$. Hence, by Eq. (4.7), we get

$$\text{C.F.} = (A_1 + A_2 t)\, e^t = (A_1 + A_2 \ln x).\, x.$$

Also, by Eq. (5.6),

$$\text{P.I.} = \{1/(D' - 1)^2\}(e^{2t}) = e^{2t}/(2 - 1)^2 = e^{2t} = x^2.$$

Hence, the solution is

$$y = \text{C.F.} + \text{P.I.} = (A_1 + A_2 \ln x).\, x + x^2$$

$\Rightarrow$

$$dy / dx = A_1 + A_2 \ln x + A_2 + 2x.$$

Applying the initial conditions, we derive $A_1 = 0$ and $A_2 = -2$. Hence, the solution reduces to $y = x^2 - 2x.\ln x$.

(iv) Putting $x + a = z$ and proceeding as in Example 7.1, the differential equation transforms into a homogeneous form:

$$(z^2.D'^{\,2} - 4\, z.\, D' + 6)\, y = z - a.$$

Applying the change of parameters as in Example 7.1, above equation further transforms to

$$\{D'' (D'' - 1) - 4\, D'' + 6\}\, y \equiv (D''^{\,2} - 5\, D'' + 6)\, y = e^{\,t} - a.$$

Its auxiliary equation $m^2 - 5m + 6 \equiv (m-2)(m-3) = 0$ has distinct roots $m = 2, 3$. Hence, by Eq. (4.3), we get

$$\text{C.F.} = A_1\, e^{\,2t} + A_3\, e^{\,3t} = A_1\, z^{\,2} + A_2\, z^3 = A_1\, (x+a)^2 + A_2\, (x+a)^3.$$

Also, by Eq. (5.6),

$$\text{P.I.} = \{1/ (D'' - 2)(D'' - 3)\}\, e^t - \{1/ (D''^2 - 5D'' + 6)\,(a)$$

$$= e^{\,t}/ 2 - (1/6)\, \{\, 1 - (5D'' - D''^2)/ 6\, \}^{-1}\, (a)$$

$$= e^{\,t}/ 2 - (1/6)\, \{1 + (5D'' - D''^2)/ 6 + \dots\}\, (a)$$

$$= e^{\,t}/ 2 - a / 6 = (x + a) / 2 - a / 6 = x / 2 + a / 3.$$

Hence, the solution is

$$y = \text{C.F.} + \text{P.I.} = A_1\, (x+a)^2 + A_2\, (x+a)^3 + x / 2.\ ]$$

10.15. Solve the following simultaneous differential equations:

(i) $dx / dt + 2dy / dt + x + 7y = e^t - 3,\quad dy / dt - 2x + 3y = 12 - 3e^t,$

(ii) $d^2x/dt^2 + m^2 y = 0,\qquad d^2y/dt^2 - m^2 x = 0,$

(iii) $d^2x/dt^2 + 4x + y = t.e^t,\ d^2y/dt^2 + y - 2x = \sin^2 t,$

(iv) $d^2x/dt^2 - 2\,dy/dt - x = e^t.\cos t,\ d^2y/dt^2 + 2dx/dt - y = e^t.\sin t,$

(v) $dx/(y+z) = dy/(z+x) = dz/(x+y),$

(vi) $dx/\cos(x+y) = dy/\sin(x+y) = dz/z,$

[**Hint: (i)** Denoting the differential operator d/dt by D and rewriting the equations:

$$(D+1)x + (2D+7)y = e^t - 3, \qquad (10.3)$$

$$-2x + (D+3)y = -3e^t + 12. \qquad (10.4)$$

Eliminating x from these by computing 2.(10.3) + $(D+1)$ (10.4), we get

$$(D^2 + 8D + 17)\,y = 2(e^t - 3) + (D+1)(-3e^t + 12) = 6 - 4e^t. \quad (10.5)$$

Its auxiliary equation $m^2 + 8m + 17 = 0$ has complex roots $m = -4 \pm i$. Hence, by (4.8b), we have

$$\text{C.F.} = (A_1 \cos t + A_2 \sin t)\, e^{-4t}.$$

and, by Note 5.1 and Eq. (5.6),

$$\text{P.I.} = \{1/(D^2 + 8D + 17)\}\,(6 - 4e^t) = -4e^t/26$$

$$+ (1/17)\,\{1 + (8D + D^2)/17\}^{-1}(6)$$

$$= -2e^t/13 + (1/17)\,\{1 - (8D + D^2)/17 + \ldots\}(6) = -2e^t/13 + 6/17.$$

Hence, Eq. (10.5) has solution

$$y = \text{C.F.} + \text{P.I.} = (A_1 \cos t + A_2 \sin t)\, e^{-4t} - 2e^t/13 + 6/17$$

$\Rightarrow$

$$(D+3)\,y = \{-A_1 \sin t + A_2 \cos t - (A_1 \cos t + A_2 \sin t)\}.\, e^{-4t}$$

$$- 8.\,e^t/13 + 18/17.$$

Putting for these values in Eq. (10.4), there results

$$x = (1/2)\{(A_2 - A_1)\cos t - (A_1 + A_2)\sin t\}.e^{-4t} + 31(e^{t}/26 - 3/17).$$

The values of x and y constitute a general solution of the given differential equations.

(ii) Rewriting the equations: $D^2 x + m^2 y = 0$, $-m^2 x + D^2 y = 0$; and eliminating x from these, we get

$$(D^4 + m^4)\,y = 0.$$

Its auxiliary equation $M^4 = -m^4 = \{\cos(2n+1)\pi + i\sin(2n+1)\pi\}m^4$, n being an integer, has roots

$$M = m\,\{\cos\{(2n+1)\pi/4\} + i\sin\{(2n+1)\pi/4\},$$

by De Moivre's theorem. Giving four different values to $n = 0, 1, 2, 3$, the roots are

$$M_1 = m\,(\cos\pi/4 + i\sin\pi/4) = m\,(1+i)/\sqrt{2},$$

$$M_2 = m\,\{\cos(3\pi/4) + i\sin(3\pi/4)\} = m\,(-1+i)/\sqrt{2},$$

$$M_3 = m\,\{\cos(5\pi/4) + i\sin(5\pi/4)\} = -m\,(1+i)/\sqrt{2},$$

$$M_4 = m\,\{\cos(7\pi/4) + i\sin(7\pi/4)\} = m\,(1-i)/\sqrt{2},$$

i.e. $m(1 \pm i)/\sqrt{2}$ and $(-1 \pm i)/\sqrt{2}$. Hence, by Eq. (4.8b), the solution is

$$y = \text{C.F.} = A_1 e^{mt/\sqrt{2}}\cos(mt/\sqrt{2} + A_2) + A_3 e^{-mt/\sqrt{2}}\cos(mt/\sqrt{2} + A_4)$$

$\Rightarrow$

$$Dy = (m/\sqrt{2})\,[A_1 e^{mt/\sqrt{2}}\{\cos(mt/\sqrt{2} + A_2) - \sin(mt/\sqrt{2} + A_2)\}$$

$$- A_3 e^{-mt/\sqrt{2}}\{\cos(mt/\sqrt{2} + A_4) + \sin(mt/\sqrt{2} + A_4)\}],$$

and

$$D^2 y = m^2\{-A_1 e^{mt/\sqrt{2}}.\sin(mt/\sqrt{2} + A_2) + A_3 e^{-mt/\sqrt{2}}.\sin(mt/\sqrt{2} + A_4)\}.$$

Putting for these values, the second given differential equation determines

$$x = -A_1 e^{mt/\sqrt{2}}.\sin(mt/\sqrt{2} + A_2) + A_3 e^{-mt/\sqrt{2}}.\sin(mt/\sqrt{2} + A_4)\}.$$

(iii) Rewriting the equations:

$$(D^2 + 4)\, x + y = t.\, e^t, \qquad -2x + (D^2 + 1)\, y = \sin^2 t,$$

and eliminating x from these, we get

$$(D^4 + 5D^2 + 6)\, y = 2t.e^{\,t} + (D^2 + 4) \sin^2 t = 2\,(t.\, e^{\,t} + 1). \quad (10.6)$$

Its auxiliary equation

$$m^4 + 5m^2 + 6 \equiv (m^2 + 2)\,(m^2 + 3) = 0$$

has roots $m = \pm \sqrt{2}\, i$ and $\pm \sqrt{3}\, i$. Hence, by Eq. (4.8c), we get

$$\text{C.F.} = A_1 \cos(\sqrt{2}\, t + A_2) + A_3 \cos(\sqrt{3}\, t + A_4);$$

and, by Eq. (5.14),

$$\text{P.I.} = (D^4 + 5D^2 + 6)^{-1}\,(2t.\, e^{\,t} + 2)$$

$$= 2e^{\,t}\,\{(D + 1)^4 + 5(D + 1)^2 + 6\}^{-1}\,(t) + (D^4 + 5D^2 + 6)^{-1}\,(2)$$

$$= 2e^{\,t}\,\{(D^4 + 4D^3 + 11D^2 + 14D + 12\}^{-1}\,(t) + \{1 + (5D^2 + D^4)/6\}^{-1}\,(1/3)$$

$$= (e^{\,t}/6)\,(1 + 7D/6 + 11D^2/12 + \ldots)^{-1}\,(t) + \{1 - 5D^2/6 - \ldots\}\,(1/3)$$

$$= (e^{\,t}/6)\,(1 - 7D/6 - \ldots)(t) + 1/3 = (e^{\,t}/6)\,(t - 7/6) + 1/3.$$

Therefore, solution of Eq. (10.6) is

$$y = \text{C.F.} + \text{P.I.}$$

$$= A_1 \cos(\sqrt{2}\, t + A_2) + A_3 \cos(\sqrt{3}\, t + A_4) + (6t - 7).e^{\,t}/36 + 1/3$$

$$D\,y = -\,\{A_1 \sqrt{2} \sin(\sqrt{2}t + A_2) + A_3\sqrt{3} \sin(\sqrt{3}t + A_4)\} + (6t - 1).e^{\,t}/36$$

$\Rightarrow$

$$D^2\, y = -\,\{2A_1 \cos(\sqrt{2}t + A_2) + 3A_3 \cos(\sqrt{3}t + A_4)\} + (6t + 5).e^{\,t}/36,$$

and

$$2\,x = (D^2 + 1)\, y - \sin^2 t$$

$$= -\,\{A_1 \cos(\sqrt{2}t + A_2) + 2A_3 \cos(\sqrt{3}t + A_4)\} + (6t - 1).e^{\,t}/18 + 1/3 - \sin^2 t$$

$\Rightarrow$

$$x = -\,\{(A_1/2) \cos(\sqrt{2}t + A_2) + A_3 \cos(\sqrt{3}t + A_4)\}$$

$$+ (6t - 1).\, e^{t}/36 + 1/6 - (1/2)\sin^2 t.$$

(iv) Rewriting the equations:

$$(D^2 - 1)\,x - 2D\,y = e^{t}.\cos t, \qquad 2Dx + (D^2 - 1)\,y = e^{t}.\sin t,$$

and eliminating y, we get

$$\{(D^2 - 1)^2 + 4D^2\}\,x = (D^2 - 1)\,(e^{t}.\cos t) + 2D\,(e^{t}.\sin t),$$

or

$$(D^2 + 1)^2\,x = e^{t}.\cos t. \qquad (10.7)$$

Its auxiliary equation $(m^2 + 1)^2 = 0$ has repeated complex roots $m = \pm\, i$. Hence, by Eq. (4.9), we get

$$\text{C.F.} = (A_1 + A_2 t)\cos t + (A_3 + A_4 t)\sin t.$$

Also, by Eqs. (5.8) and (5.14),

$$\text{P.I.} = \{1/(D^2 + 1)^2\}\,(e^{t}.\cos t) = e^{t}\{(D + 1)^2 + 1\}^{-2}\,(\cos t)$$

$$= e^{t}\,(D^2 + 2D + 2)^{-2}\,(\cos t) = e^{t}\,(2D + 1)^{-2}\,(\cos t), \text{ by Eq. (5.8)}$$

$$= e^{t}\,(1 + 4D + 4D^2)^{-1}\,(\cos t) = e^{t}\,(4D - 3)^{-1}\,(\cos t), \text{ by Eq. (5.8)}$$

$$= (e^{t}/4)\,(D - 3/4)^{-1}\,(\cos t) = (e^{t}/4).\, e^{3t/4}.\int e^{-3t/4}.\cos t.\,dt, \text{ by Eq. (5.4)}$$

$$= (e^{t}/4)\,\{(-3/4)\cos t + \sin t\}\,/\,(9/16 + 1) = (-3\cos t + 4\sin t)\,e^{t}/25,$$

by Eq. (5.19). Therefore, solution of Eq. (10.7) is

$$y = \text{C.F.} + \text{P.I.}$$

$$= (A_1 + A_2 t)\cos t + (A_3 + A_4 t)\sin t + (4\sin t - 3\cos t)\,e^{t}/25.$$

On the other hand, eliminating the terms containing x from the given differential equations, we get

$$\{(D^2 - 1)^2 + 4D^2\}\,y = (D^2 - 1)\,(e^{t}.\sin t) - 2D\,(e^{t}.\cos t),$$

or

$$(D^2 + 1)^2\,y = e^{t}.\sin t. \qquad (10.8)$$

Comparing Eqs. (10.7) and (10.8), it may be noted that the independent variable t is now replaced by $t + \pi/2$. Hence, the C.F. of Eq. (10.8) may now be taken as

$$\text{C.F.} = (A_1 + A_2 t)\cos(t + \pi/2) + (A_3 + A_4 t)\sin(t + \pi/2).$$

The particular integral may be evaluated similarly:

$$\text{P.I.} = (e^{t}/4)\,(D - 3/4)^{-1}(\sin t) = (e^{t}/4).e^{3t/4}.\int e^{-3t/4}.\sin t.\,dt$$

$$= -(e^{t}/4)\,\{(3/4)\sin t + \cos t\}/\,(9/16 + 1) = -(3\sin t + 4\cos t)\,e^{t}/25,$$

by Eq. (5.17). Therefore, solution of Eq. (10.8) is

$$y = \text{C.F.} + \text{P.I.}$$

$$= -(A_1 + A_2 t)\sin t + (A_3 + A_4 t)\cos t - (3\sin t + 4\cos t)\,e^{t}/25.$$

(v) Given fractions are also equal to

$$(dx - dy)\,/\,(x - y) = (dy - dz)\,/\,(y - z) = (dz - dx)\,/\,(z - x), \quad (10.9)$$

which is the same differential eqn. as Eq.(8.18). However, taking the first and the last fraction in above equations and integrating the either sides w.r.t. the respective variables, there also results

$$\ln(x - y) = \ln(z - x) + \ln a \quad \Rightarrow \quad x - y = a\,(z - x).$$

On the other hand, taking the last two fractions in Eq. (10.9) (as in Example 8.5) yields the soln. $y - z = b\,(z - x)$. Thus, the desired soln. is a family of curves of intersection of these surfaces, a and b being arbitrary constants of integration.

(vi) First two fractions yield

$$dy\,/\,dx = \tan(x + y) = t\ \text{(say)}, \quad (10.10)$$

$\Rightarrow$

$$x + y = \tan^{-1} t. \quad (10.11)$$

Differentiating Eq. (10.11) w.r.t. x and putting from Eq. (10.10), there results

$$1 + dy / dx \equiv 1 + t = \{1 / (1 + t^2)\} (dt / dx)$$

$\Rightarrow$

$$dt / (1 + t) (1 + t^2) = dx, \qquad (10.12)$$

or

$$\{1 / (1 + t) + (1 - t) / (1 + t^2)\} dt = 2 dx$$

with integral

$$\ln (1 + t) + \tan^{-1} t - (1/2) \{ \ln (1 + t^2) \}$$

$$\equiv \ln \{(1 + t) / \sqrt{(1 + t^2)}\} + \tan^{-1} t = 2x + \ln a.$$

For Eq. (10.10),

$$(1 + t) / \sqrt{(1 + t^2)} = \{1 + \tan (x + y)\} / \sqrt{\{1 + \tan^2 (x + y)\}}$$

$$= \cos (x + y) + \sin (x + y),$$

and $\tan^{-1} t - 2x = y - x$. Therefore, above integral simplifies to

$$\ln \{\cos (x + y) + \sin (x + y)\} + y - x = \ln a,$$

or,

$$\{\cos (x + y) + \sin (x + y)\} . e^{y - x} = a.$$

On the other hand, the first and the last fractions in the given differential equation yield

$$dz / z = \sec (x + y) dx = \sqrt{(1 + t^2)} dx = dt / (1 + t) \sqrt{(1 + t^2)},$$

by Eqs. (10.10) and (10.12). Putting $1 + t = 1/u$ so that $dt = - du/u^2$ and

$$\sqrt{(1 + t^2)} = \sqrt{\{1 + (1/u - 1)^2\}} = \sqrt{(2u^2 - 2u + 1)} / u,$$

above equation reduces to

$$dz / z = - du / \sqrt{(2u^2 - 2u + 1)\}} \Rightarrow \sqrt{2} dz/z = - du / \sqrt{\{(u - 1/2)^2 + 1/4}.$$

Integrating it term wise w.r.t. respective variables, we get

$$\sqrt{2} . \ln z = \ln b - \ln [(u - 1/2) + \sqrt{\{(u - 1/2)^2 + 1/4\}}]$$

$$= \ln b - \ln [(2u - 1) + \sqrt{\{(2u - 1)^2 + 1\}}] + \ln 2.$$

The constant term ln 2 may be adjusted with the arbitrary constant of integration ln b. Since

$$u = 1/(1+t)$$

⇒

$$2u - 1 = (1-t)/(1+t) = \{1 - \tan(x+y)\} / \{1 + \tan(x+y)\}$$

$$= \{\cos(x+y) - \sin(x+y)\} / \{\cos(x+y) + \sin(x+y)\},$$

and

$$\sqrt{\{(2u-1)^2 + 1\}} = \sqrt{\{2(1+t^2)\}} / (1+t)$$

$$= \sqrt{2}\sec(x+y) / \{1 + \tan(x+y)\} = \sqrt{2} / \{\cos(x+y) + \sin(x+y)\}$$

⇒

$$(2u - 1) + \sqrt{\{(2u-1)^2 + 1\}}$$

$$= \{\cos(x+y) - \sin(x+y) + \sqrt{2}\} / \{\cos(x+y) + \sin(x+y)\}$$

$$= \{\cos(x+y).\cos(\pi/4) - \sin(x+y).\sin(\pi/4) + 1\}$$

$$/ \{\cos(x+y).\sin\pi/4 + \sin(x+y).\cos\pi/4\}$$

$$= \{\cos(x+y+\pi/4) + 1\} / \sin(x+y+\pi/4)$$

$$= \cos\{(x+y)/2 + \pi/8\} / \sin\{(x+y)/2 + \pi/8\} = \cot\{(x+y)/2 + \pi/8\},$$

above integral reduces to

$$\sqrt{2}.\ln z = \ln b - \ln\cot\{(x+y)/2 + \pi/8\}$$

⇒

$$z^{\sqrt{2}}.\cot\{(x+y)/2 + \pi/8\} = b. //$$

10.16. Find the surface generated by the line $dx/x = dy/y = dz/z$ that intersects the helix

$$x^2 + y^2 = r^2, \qquad z = k.\tan^{-1}(y/x).$$

[Hint: Taking two fractions at a time and integrating the equations so obtained, with respect to the respective variables, there result

$$\ln x = \ln y + \ln a \Rightarrow x = a\,y, \tag{10.13}$$

and

$$y = b\,z. \tag{10.14}$$

Thus, the surface is a family of cylinders generated by these lines. Elimination of y from these equations also implies

$$x = a\, b\, z \qquad \Rightarrow \qquad x^2 = a^2 b^2 z^2. \qquad (10.15)$$

As the surface intersects the helix, substitutions from Eqs. (10.13) and (10.14) in the equation of helix yield

$$a^2 y^2 + y^2 = r^2 \qquad \Rightarrow \qquad a^2 = (r^2 - y^2) / y^2,$$

and

$$x^2 + b^2 z^2 = r^2 \qquad \Rightarrow \qquad b^2 = (r^2 - x^2) / z^2.$$

Elimination of a^2, b^2 from Eq. (10.15), determines a unique surface

$$x^2 = (r^2 - y^2)\,(r^2 - x^2)\,z^2 / y^2 z^2$$

$\Rightarrow$

$$x^2 y^2 z^2 = \{r^4 - r^2 (x^2 + y^2) + x^2 y^2\}\, z^2,$$

$\Rightarrow$

$$r^2 (x^2 + y^2)\, z^2 = r^4 z^2,$$

or

$$(x^2 + y^2).\ \{k.\ \tan^{-1} (y / x)\}^2 = r^2 z^2.\ ]$$

CHAPTER 3

BILINEAR (OR *MÖBIUS*) TRANSFORMATIONS

§ 1. Introduction

The transformation

$$w = (a\,z + b) / (c\,z + d), \tag{1.1}$$

where a, b, c, d are some complex constants satisfying

$$ad - bc \neq 0, \tag{1.2}$$

is called a *bilinear* transformation.

Note 1.1. The transformation (1.1) was first studied by August Ferdinand *Möbius* (1790 – 1868 A.D.) hence is also called after him.

The condition (1.2) ensures that

$$dw / dz = \{a\,(c\,z + d) - c\,(a\,z + b)\} / (c\,z + d\,)^2$$

$$= (ad - bc) / (c\,z + d\,)^2 \neq 0,$$

so that the transformation becomes conformal. In the lack of above condition, it causes every point of the z–plane becoming a critical point.

Theorem 1.1. The inverse mapping of (1.1) is also bilinear.

Proof. Solving (1.1) for z (in terms of w):

$$(cw - a)\,z = b - w\,d \;\Rightarrow\; z = \{(-d)\,w + b\} / (c\,w - a). \tag{1.3}$$

Comparing (1.3) with (1.1) we have the statement. Further,

$$dz\,/dw = \{-(c\,w - a)\,d \;-\; c\,(-w\,d + b)\} / (c\,w - a)^2$$

$$= (ad - bc) / (c\,w - a)^2 \neq 0,$$

in view of the condition (1.2) making (1.3) also conformal. //

Note 1.2. It is clear from (1.1) that each point of z–plane except $z = -d/c$ corresponds to a unique point in the w–plane. Similarly, it is also evident from (1.3) that each point of w–plane except $w = a/c$ maps onto a unique point in z–plane.

Furthermore, the exceptional points $z = -d/c$ of z–plane (respectively $w = a/c$ of w–plane) map onto infinity of w–plane (resp. z–plane).

1.1. Invariant points of a bilinear transformation

If a point $z = z_1$ (of z–plane) maps onto itself in the w–plane, i.e. $w = z$, then (1.1) becomes

$$z = (az+b)/(cz+d) \Rightarrow cz^2+(d-a)z-b = 0. \qquad (1.4)$$

The roots of this quadratic equation are called the *invariant* or *fixed* points of the bilinear transformation (1.1). If these roots are equal the transformation is called *parabolic.*

Dividing both the numerator and denominator in the right hand side of Eq. (1.1) by one of the non-vanishing constant coefficient (say b), and rewriting it as

$$w = (a_1 z+1)/(c_1 z+d_1), \qquad (1.5)$$

where $a_1 = a/b$, etc. we note that there are only three essential constants. Thus, only three conditions are required to determine a bilinear transformation. Consequently, three distinct points z_1, z_2, z_3 (of z–plane) can be mapped onto any three specified points w_1, w_2, w_3 (of w–plane).

Theorem 1.2. A bilinear transformation maps circles into circles.

Proof. Rewriting (1.1) as

$$w = (a/c).\, cz/(cz+d)+b/(cz+d)$$

$$= (a/c).\, (cz+d-d)/(cz+d)+b/(cz+d)$$

$$= a/c+(bc-ad)/c\,(cz+d) = a/c+(bc-ad)/c^2\,(z+d/c)$$

$$= a/c+(bc-ad)/c^2\,w_1 = a/c+\{(bc-ad)/c^2\}.\,w_2 = a/c+w_3, \qquad (1.6)$$

where $\quad w_1 = z + d / c, \;\; w_2 = 1/ w_1, \;\; w_3 = \{(bc - ad) / c^2\} w_2 .$

Thus, the bilinear transformation (1.1) is the combination of one or other of the standard (conformal) transformations. It is known that each of those standard transformations map circles onto circles. Hence, their combination (1.6) also does the same.

Theorem 1.3. A bilinear transformation preserves the cross–ratio of four points.

Proof. Let the four points $z_1, \; z_2, \; z_3, z_4$ (of z–plane) map onto the respective points $w_1, \; w_2, \; w_3, \; w_4$ (of w–plane) under the bilinear transformation (1.1). If these points are finite, we have

$$w_i - w_j = (a z_i + b) / (c z_i + d) - (a z_j + b) / (c z_j + d)$$

$$= (ad z_i + bc z_j - bc z_i - ad z_j) / (c z_i + d)(c z_j + d)$$

$$= (ad - bc)(z_i - z_j) / (c z_i + d)(c z_j + d).$$

Choosing $i, j = 1, 2, 3, 4$ above relation gives

$$(w_1 - w_2)(w_3 - w_4)$$

$$= (ad - bc)^2 (z_1 - z_2)(z_3 - z_4) / (cz_1 + d)(cz_2 + d)(cz_3 + d)(cz_4 + d),$$

and

$$(w_1 - w_4)(w_3 - w_2)$$

$$= (ad - bc)^2 (z_1 - z_4)(z_3 - z_2) / (cz_1 + d)(cz_4 + d)(cz_3 + d)(cz_2 + d),$$

making the cross–ratio of four points an invariant under (1.1):

$$(w_1 - w_2)(w_3 - w_4) / (w_1 - w_4)(w_3 - w_2)$$

$$= (z_1 - z_2)(z_3 - z_4) / (z_1 - z_4)(z_3 - z_2). \; // \qquad (1.7)$$

Example 1.1. Find the bilinear transformation mapping the points $z = 1, \; i, -1$ (of z–plane) onto the points $w = i, 0, -i$. Hence, find the image of $z < 1$ and the invariant points of the transformation.

Solution. (i) Let a transformation (1.1) map the points

$$z_1 = 1, \quad z_2 = i, \quad z_3 = -1 \quad \text{and} \quad z_4 = z, \quad \text{(of } z\text{–plane)}$$

onto

$$w_1 = i, \quad w_2 = 0, \quad w_3 = -i \quad \text{and} \quad w_4 = w, \quad \text{(of } w\text{–plane)}$$

respectively. Applying (1.7) we then have

$$i\,(-i-w)\,/\,(i-w)\,(-i) = (1-i)\,(-1-z)\,/\,(1-z)\,(-1-i)$$

$\Rightarrow$

$$(w+i)\,/\,(w-i) = (1-i)\,(1+z)\,/\,(1+i)\,(z-1)$$

$$= \{\,(1-i)\,z + (1-i)\,\}\,/\,\{\,(1+i)\,z - (1+i)\,\}.$$

Applying the method of *componendo* and *dividendo* we derive

$$w\,/\,i = (z-i)\,/\,(-i\,z+1) \quad \Rightarrow \quad w = (i\,z+1)\,/\,(-i\,z+1). \quad (1.8)$$

(ii) Solving (1.8) for z: $\quad i\,z\,(1+w) = w-1$

$\Rightarrow$

$$z = (w-1)\,/\,i\,(w+1) = i\,(1-w)\,/\,(1+w). \qquad (1.9)$$

the inequality $|\,z\,| < 1$ reduces to

$$|\,z\,| = |\,1-w\,|\,/\,|\,1+w\,| < 1 \;\Rightarrow\; |\,1-u-iv\,| < |\,1+u+iv\,|,$$

or

$$(1-u)^2 + v^2 < (1+u)^2 + v^2 \;\Rightarrow\; 0 < 4u \;\Rightarrow\; u > 0.$$

Thus, the interior of the circle $|\,z\,| = 1$ transforms onto the entire half of the w–plane to the right of the imaginary axis.

(iii) To find the invariant points of the transformation, we put $w = z$ in (1.8):

$$i\,z^2 + (i-1)\,z + 1 = 0$$

$\Rightarrow$

$$z = [1-i \pm \sqrt{\{(i-1)^2 - 4i\}}]\,/\,2i = -\{1+i \mp \sqrt{(6i)}\}\,/\,2$$

determining the invariant points. //

Example 1.2. Show that the transformation (1.1) transforms the circle $|\,w\,| = 1$ onto a straight line in z–plane if $|\,a\,| = |\,c\,|$.

Solution. From (1.1) we get

$$|w| = |az+b| / |cz+d| = 1 \;\Rightarrow\; |a(x+iy)+b| = |c(x+iy)+d|$$

$$\Rightarrow \qquad (ax+b)^2 + (ay)^2 = (cx+d)^2 + (cy)^2,$$

or,

$$a^2(x^2+y^2) + 2abx + b^2 = c^2(x^2+y^2) + 2(cd)x + d^2.$$

It represents a straight line if the second degree terms on either side get cancelled, i.e. when $a^2 = c^2 \Rightarrow |a| = |c|$. //

Example 1.3. Find the bilinear transformation that maps the points $z = 1, i, -1$ (of z–plane) onto the points $w = 2, i, -2$ (of w–plane). Also, find the fixed and critical points of the transformation.

Solution. (i) Let the transformation (1.1) map the points

$$z_1 = 1,\; z_2 = i,\; z_3 = -1,\; z_4 = z \;\text{ onto }\; w_1 = 2,\; w_2 = i,\; w_3 = -2,\; w_4 = w$$

respectively. Applying the rule (1.7), we derive

$$(2-i)(-2-w)/(2-w)(-2-i) = (1-i)(-1-z)/(1-z)(-1-i)$$

$$\Rightarrow \; (-2+i)(w+2)/(2+i)(w-2) = (i-1)(z+1)/(i+1)(z-1).$$

Applying method of componendo and dividendo, we derive

$$(iw-4)/2(-w+i) = (iz-1)/(-z+i)$$

$$\Rightarrow$$

$$w = (2i-6z)/(iz-3). \tag{1.10}$$

(ii) For fixed points we put $w = z$ in above equation:

$$z(iz-3) = (2i-6z) \quad \Rightarrow \quad iz^2 + 3z - 2i = 0,$$

or,

$$z^2 - 3iz - 2 = 0 \;\Rightarrow\; z = \{3i \pm \sqrt{(-1)}\}/2 = 2i,\; i.$$

(iii) Differentiating (1.10) with respect to z we note that

$$dw/dz = \{-6(iz-3) + (6z-2i)\,i\}/(iz-3)^2 = 20/(iz-3)^2 \neq 0.$$

Hence, there are no critical points of the transformation. //

CHAPTER 4

MULTIVALUED FUNCTIONS

§ 1. History

The practice of allowing *function* in mathematics to mean also *multivalued function* dropped out of usage at some point in the first half of the twentieth century. Some evolution can be seen in different editions of *A Course of Pure Mathematics* by G.H. Hardy, for example. It probably persisted longest in the theory of special functions, for its occasional convenience.

The theory of multivalued functions was fairly systematically developed for the first time in Claude Berge's *Topological spaces* (1963).

§ 2. Introduction

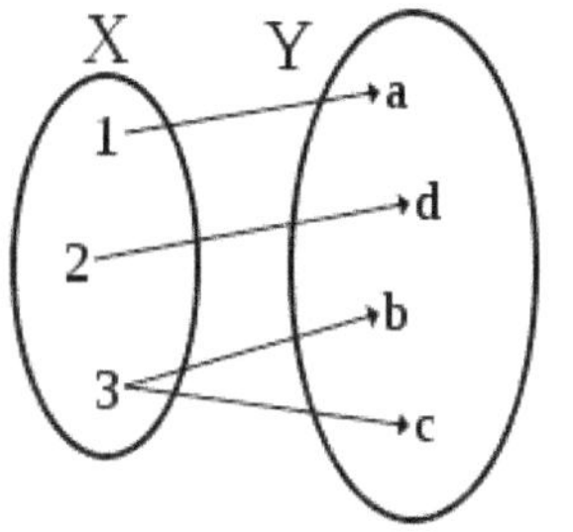

In mathematics, a *multivalued function* from a domain X to a codomain Y is, roughly speaking, a function which may associate more than one element of Y to each element of X. Formally, a multivalued function is an ordinary function from X to the set of the non–empty subsets of Y, or a left–total relation. In this context, an ordinary function is often called a *single–valued function* to avoid confusion.

Note 2.1. Above diagram does not represent a *true* function, because the element 3 in X is associated with two elements, b and c, in Y.

The term *multi-valued function* originated from analytic continuation in complex analysis. It often occurs that one knows the value of a complex analytic function $f(z)$ in some neighbourhood of a point $z = a$. This is the case for functions defined by the implicit function theorem or by a Taylor series around $z = a$. In such a situation, one may extend the domain of the single–valued function $f(z)$ along curves in the complex plane starting at a. In doing so, one finds that the value of the extended function at a point $z = b$ depends on the chosen curve from a to b; since none of the new values is more natural than the others, all of them are incorporated into a multivalued function. For example, let $f(z) = \sqrt{z}$ be the usual square root function on positive real numbers. One may extend its domain to a neighbourhood of $z = 1$ in the complex plane, and then further along curves starting at $z = 1$, so that the values along a given curve vary continuously from $\sqrt{1} = 1$. Extending to negative real numbers, one gets two opposite values of the square root such as $\sqrt{(-1)} = \pm i$, depending on whether the domain has been extended through the upper or the lower half of the complex plane. This phenomenon is very frequent, occurring for n^{th} roots, logarithms and inverse trigonometric functions.

To define a single–valued function from a complex multivalued function, one may distinguish one of the multiple values as the principal value, producing a single–valued function on the whole plane which is discontinuous along certain boundary curves. Alternatively, dealing with the multivalued function allows having something that is everywhere continuous, at the cost of possible value changes when one follows a closed path (monodromy). These problems are resolved in the theory of Riemann surfaces: to consider a multivalued function $f(z)$ as an ordinary function without discarding any values, one multiplies the domain into a many–layered covering space, a manifold which is the Riemann surface associated to $f(z)$.

Example 2.1. Every real number greater than zero has two real square roots, so that square root may be considered a multivalued function. For example, we may write $\sqrt{4} = \pm 2 = \{2, -2\}$, whereas zero has only one square root, $\sqrt{0} = 0$.

Example 2.2. Each non-zero complex number has *two* square roots, *three* cube roots, and in general n, n^{th} roots. All n^{th} roots of 0 coincide with 0 itself.

Example 2.3. The complex logarithm function is multiple–valued. The values assumed by $\log_e (a + ib)$ for real numbers a and b are

$$\ln \sqrt{(a^2 + b^2)} + i \arg (a + ib) + 2n\pi i, \qquad \text{for all integers } n.$$

Example 2.4. Inverse trigonometric functions are multiple–valued because trigonometric functions are periodic. We have

$$\tan (\pi/4) = \tan (5\pi/4) = \tan (-3\pi/4) = \tan (n\pi + \pi/4) = \ldots = 1, \quad n \in \mathbb{Z}.$$

As a consequence, $\tan^{-1}$ is intuitively related to several values: $\pi/4$,

$5\pi/4$, $-3\pi/4$, and so on. We can treat $\tan^{-1}$ as a single–valued function by restricting the domain of $\tan x$ to $-\pi/2 < x < \pi/2$, a domain over which $\tan x$ is monotonically increasing. Thus, the range of $\tan^{-1} x$ becomes $-\pi/2 < y < \pi/2$. These values from a restricted domain are called *principal values*.

Example 2.5. The indefinite integral can be considered as a multivalued function. The indefinite integral of a function is the set of functions whose derivative is that function. The constant of integration follows from the fact that the derivative of a constant function is 0.

Example 2.6. The arg max is multivalued, for example $\arg\max_{x \in R} \cos x = \{2k\pi$, such that $k \in Z\}$.

All above examples of multivalued functions come about from non–injective functions. Since the original functions do not preserve all the information of their inputs, they are not reversible. Often, the restriction of a multivalued function is a partial inverse of the original function.

Multivalued functions of a complex variable have branch points. For example, for the nth root and logarithm functions, 0 is a branch point; for the arc tangent function, the imaginary units i and $-i$ are branch points. Using the branch points, these functions may be redefined to be single–valued functions, by restricting the range. A suitable interval may be found through use of a branch cut, a kind of curve that connects pairs of branch points, thus reducing the multilayered Riemann surface of the function to a single layer. As in the case with real functions, the restricted range may be called the *principal branch* of the function.

§ 3. Set–valued Analysis

It is the study of sets in the spirit of mathematical analysis and general topology. Instead of considering collections of only points, set–valued analysis considers collections of sets. If a collection of sets is endowed with a topology, or inherits an appropriate topology from an underlying topological space, then the convergence of sets can be studied.

Much of set–valued analysis arose through the study of mathematical economics and optimal control, partly as a generalization of convex analysis; the term *variational analysis* is used by RT. Rockfeller and Roger Wets, Jon Borwein and Adrian Lewis, and Boris Mordukhovich. In optimization theory, the convergence of approximating sub–differentials to a sub-differential is important in understanding necessary or sufficient conditions for any minimizing point.

There exist set–valued extensions of the following concepts from point–valued analysis: continuity, differentiation, integration, implicit function theorem, contraction maps, measure theory, fixed–point theorems, optimization, and topological theory. Equations are generalized to inclusions.

§ 4. Types of Multivalued Functions

One can differentiate many continuity concepts, primarily closed graph property and upper and lower hemicontinuity. (One should be warned that often the terms upper and lower semi-continuous are used instead of upper and lower hemi-continuous reserved for the case of weak topology in domain; yet we arrive at the collision with the reserved names for upper and lower semi continuous real–valued function). There exist also various definitions for measurability of multifunction.

§ 5. Applications

Multi-functions arise in optimal control theory, especially differential inclusions and related subjects as game theory, where the Kakutani fixed point theorem for multi-functions has been applied to prove existence of Nash equilibrium. It may be noted that in the context of game theory, a multivalued function is usually referred to as a correspondence. This among many other properties loosely associated with approximability of upper hemi-continuous multi-functions via continuous functions explains why upper hemicontinuity is more preferred than lower hemicontinuity.

In physics, multivalued functions play an increasingly important role. They form the mathematical basis for Dirac's magnetic monopoles, for the theory of defects in crystals and the resulting plasticity of materials, for vortices in superfluids and superconductors, and for phase transitions in these systems, for instance melting and quark confinement. They are the origin of gauge field structures in many branches of physics.

CHAPTER 5

INTEGRAL REPRESENTATION OF FUNCTIONS

§ 1. Introduction

Integral representations of analytic functions arose in the early stages of development of function theory and mathematical analysis in general as a suitable apparatus for the explicit representation of analytic solutions of differential equations, for the investigation of the asymptotics of these solutions and for their analytic continuation. Later integral representations of analytic functions found their applications in the solution of boundary value problems of analytic function theory and singular integral equations in the study of interior and boundary properties of analytic functions of various classes, as well as in the solution of other problems of mathematical analysis.

A wide class of integral representations of analytic functions, used for obtaining and studying analytic solutions of differential equations, can be described by the general formula

$$f(z) = \int_L K(z,t)\, v(t)\, dt,$$

where $K(z, t)$ is the *kernel* of the integral representation, $\upsilon(t)$ is its density and L is a contour (or system of contours) in the complex plane in which both variables z and t vary.

As an example, an integral representation of Bessel function of the *first kind* is discussed below.

§ 2. Bessel's Equation

The ODE

$$x^2 y'' + xy' + (x^2 - \upsilon^2)\, y = 0 \tag{2.1}$$

is called the Bessel's equation of order $\upsilon \geq 0$. In the following, we discuss the series solution of above equation about the point $x = 0$. It may be noted that this point is a *regular singular point* of the ODE.

2.1. To find a series solution of (2.1)

For $x = 0$ being a regular singular point of the ODE, there exists at least one series solution of the form

$$y = \sum_{n=0}^{\infty} c_n \, x^{n+r}. \tag{2.2}$$

Differentiating it term wise w.r.t. x under the sign of summation, we get

$$y' = \sum_{n=0}^{\infty} c_n \,(n+r)\, x^{n+r-1}, \qquad y'' = \sum_{n=0}^{\infty} c_n \,(n+r)(n+r-1)x^{n+r-2},$$

So that the ODE (2.1) reduces to

$$y'' = \sum_{n=0}^{\infty} c_n \,(n+r)(n+r-1)x^{n+r} + \sum_{n=0}^{\infty} c_n \,(n+r)\, x^{n+r} + \sum_{n=0}^{\infty} c_n \, x^{n+r+2} - \upsilon^2 \sum_{n=0}^{\infty} c_n \, x^{n+r}$$

$$= c_0\,(r^2 - \upsilon^2)\, x^r + x^r.\sum_{n=1}^{\infty} c_n\{(n+r)^2 - \upsilon^2\}x^n + x^r.\sum_{n=0}^{\infty} c_n \, x^{n+2} = 0. \tag{2.3}$$

Comparing the coefficients of like terms in above identity there results

$$c_0\,(r^2 - \upsilon^2) = 0, \qquad c_n\{(n+r)^2 - \upsilon^2\} = 0, \qquad n = 1, 2, 3, \ldots$$

The first of these eqns. yields $r^2 - \upsilon^2 = 0$, called the *indicial equation* with roots $r = \pm\,\upsilon$. Thus, there arise two cases:

Case 2.1. $r = \upsilon$, the identity (2.3) reduces to

$$x^{\upsilon}.\left[\sum_{n=1}^{\infty} c_n.n\,(n+2\upsilon).x^n + \sum_{n=0}^{\infty} c_n \, x^{n+2}\right]$$

$$= x^{\upsilon}.\left[c_1\,(1+2\upsilon)x + \sum_{n=2}^{\infty} c_n.n\,(n+2\upsilon).x^n + \sum_{n=0}^{\infty} c_n \, x^{n+2}\right]$$

$$= x^{\upsilon}.\left[c_1\,(1+2\upsilon)x + \sum_{k=0}^{\infty} c_{k+2}.(k+2)\,(k+2+2\upsilon).x^{k+2} + \sum_{k=0}^{\infty} c_k \, x^{k+2}\right]$$

$$= x^{\upsilon}.\left[c_1\,(1+2\upsilon)x + \sum_{k=0}^{\infty}\{c_{k+2}.(k+2)\,(k+2+2\upsilon) + c_k\}\, x^{k+2}\right] = 0.$$

Comparing the coefficients of like terms, we thus have

$$(1 + 2\upsilon)\, c_1 = 0 \tag{2.4}$$

$$c_{k+2} = -c_k \,/\, (k+2)\,(k+2+2\upsilon) \tag{2.5}$$

where $k = 0, 1, 2, \ldots$ Taking the choice $c_1 = 0$, (2.5) determines $c_3 = c_5 = c_7 = \ldots = 0$. Further, writing $k+2 = 2n$, $n = 1, 2, 3, \ldots$, relations (2.5) may be re-written as

$$c_{2n} = -c_{2n-2} \,/\, 2n\,(2n+2\upsilon) = -c_{2n-2} \,/\, 2^2 .\, n\,(n+\upsilon), \tag{2.6}$$

determining

$$c_2 = -c_0 \,/\, 2^2 .\, 1.\, (1+\upsilon),$$

$$c_4 = -c_2 \,/\, 2^2 .\, 2.\, (2+\upsilon) = c_0 \,/\, 2^4 .\, 1.2.\, (1+\upsilon)\,(2+\upsilon),$$

$$c_6 = -c_4 \,/\, 2^2 .3.(3+\upsilon) = -c_0 \,/\, 2^6 .1.2.3.(1+\upsilon)\,(2+\upsilon)\,(3+\upsilon) \ldots$$

Thus, by induction

$$c_{2n} = (-1)^n .\, c_0 / 2^{2n} .\, n!\,(1+\upsilon)\,(2+\upsilon)\,(3+\upsilon) \ldots . (n+\upsilon). \tag{2.7}$$

Taking $c_0 = 1/\, 2^{\upsilon}\Gamma(1+\upsilon)$ and using properties of *Gamma functions*: $\Gamma(p+1) = p.\,\Gamma(p)$. Accordingly, (2.7) further reduces to

$$c_{2n} = (-1)^n \,/\, 2^{2n+\upsilon} .\, n!\,(1+\upsilon)\,(2+\upsilon)\,(3+\upsilon) \ldots (n+\upsilon)\,\Gamma(1+\upsilon)$$

$$= (-1)^n \,/\, 2^{2n+\upsilon} .\, n!\,\Gamma(1+\upsilon+n). \tag{2.8}$$

Thus, a series solution (2.2), for above case, takes the form

$$y = \sum_{n=0}^{\infty} c_{2n}\, x^{2n+\upsilon} = \sum_{n=0}^{\infty} \frac{(-1)^n}{n!\,\Gamma(1+\upsilon+n)} \cdot \left(\frac{x}{2}\right)^{2n+\upsilon} \equiv J_{\upsilon}(x). \tag{2.9}$$

The functions $J_{\upsilon}(x)$ are called the *Bessel's functions of the first kind.*

Case 2.2. Similarly for $r = -\upsilon$, we may derive the *Bessel's functions of the second kind*

$$y = \sum_{n=0}^{\infty} c_{2n}\, x^{2n-\upsilon} = \sum_{n=0}^{\infty} \frac{(-1)^n}{n!\,\Gamma(1-\upsilon+n)} \cdot \left(\frac{x}{2}\right)^{2n-\upsilon} \equiv J_{-\upsilon}(x). \quad (2.10)$$

Example 2.1. Taking a solution $y_1 = (\cos x)/x$, solve the differential equation

$$x.\, y'' + 2\, y' + x.\, y = 0. \quad (2.11)$$

Solution. Putting $x.\, y = t(x)$, so that on repeated derivations, it yields

$$x.\, y' + y = t', \qquad x.\, y'' + 2y' = t''.$$

Thus, (2.11) assumes the form

$$t'' + t = 0, \qquad \text{i.e.} \qquad (D^2 + 1)\, t = 0,$$

where $D \equiv dt/dx$. Its auxiliary equation is $m^2 + 1 = 0$ having imaginary roots $m = \pm\, i$. Hence, C.F. is given by

$$t = x.\, y = A_1\, e^{ix} + A_2\, e^{-ix}$$

$$= (A_1 + A_2) \cos x + i\,(A_1 - A_2) \sin x = B_1 \cos x + B_2 \sin x$$

$\Rightarrow$

$$y \equiv (t/x) = B_1 (\cos x)/x + B_2 (\sin x)/x.$$

Thus, the other solution is

$$y_2 = (\sin x)/x. \;//$$

CHAPTER 6

LAPLACE TRANSFORMS

§ 1. Integral Transform

Let $K(s, t)$ be some function of variables s and t and let the integral

$$\int_{-\infty}^{\infty} K(s, t)\, F(t)\, dt \tag{1.1}$$

be convergent. Above integral is called the *Integral transform* of the function $F(t)$ and is denoted by $T\{F(t)\}$ or $f(s)$:

$$T\{F(t)\} \equiv f(s) = \int_{t=-\infty}^{\infty} K(s, t)\, F(t)\, dt. \tag{1.2}$$

The function $K(s, t)$ is called the *kernel* of the transformation. The variable s is a parameter independent of t. Choosing

$$K(s, t) = 0 \text{ (when } t < 0), \; e^{-st} \text{ (when } t \geq 0), \tag{1.3}$$

the integral transform (1.2) becomes

$$T\{F(t)\} = \int_{t=-\infty}^{0-\varepsilon} 0.\, F(t) dt + \int_{t=0}^{\infty} e^{-st}.F(t)\, dt = \int_{t=0}^{\infty} e^{-st}.F(t)\, dt$$

$$\equiv f(s). \tag{1.4}$$

This is the special form of (1.2) called the *Laplace transform* of $F(t)$ and is also denoted by £$\{F(t)\}$. Thus, we have

$$\text{£}\{F(t)\} = \int_{t=0}^{\infty} e^{-st}.\, F(t)\, dt \equiv f(s). \tag{1.5}$$

Note 1.1. The existence of the Laplace transform depends upon the convergence of above integral.

§ 2. Laplace Transform of Some Standard Functions

2.1. $\text{£}(1) = \int_{t=0}^{\infty} e^{-st}\, dt = \left[-e^{-st}/s\right]_0^{\infty} = 1/s$, when $s > 0$. (2.1)

$$\mathbf{2.2.}\ \pounds\,(t) = \int_{t=0}^{\infty} t.\,e^{-st}\,dt = \left[-t.\,e^{-st}/s\right]_0^{\infty} + \int_{t=0}^{\infty} (e^{-st}/s)\,dt$$

$$= \lim_{t\to\infty} (-t/s.\,e^{st}) + \left[-e^{-st}/s^2\right]_0^{\infty}$$

$$= \lim_{t\to\infty} (-1/s^2.\,e^{st}) + 1/s^2 = 1/s^2, \qquad (2.2)$$

where, s > 0 and L'Hospital's rule are used to evaluate the limit.

$$\mathbf{2.3.}\quad \pounds\,(t^n) = \int_{t=0}^{\infty} t^n.\,e^{-st}\,dt = \int_{t=0}^{\infty} (u^n/s^n).\,e^{-u}\,du/s,$$

where we have put $st = u$ so that $t = u/s$ and $dt = (du)/s$. Applying the definition of Gamma function, above integral reduces to

$$\pounds\,(t^n) = (1/s^{n+1}) \int_{t=0}^{\infty} u^n.\,e^{-u}\,du = \Gamma(n+1)/s^{n+1} = (n!)/s^{n+1}, \quad (2.3)$$

for positive integer n.

$$\mathbf{2.4.}\qquad \pounds\,(e^{at}) = \int_{t=0}^{\infty} e^{at}.\,e^{-st}\,dt = \int_{t=0}^{\infty} e^{-(s-a)t}\,dt$$

$$= \left[-e^{-(s-a)t}/(s-a)\right]_0^{\infty} = 1/(s-a), \qquad (2.4)$$

when $s > a$.

2.5. Trigonometric functions: To evaluate the Laplace of *sine* and *cosine* functions we consider the Laplace transform of the exponential function e^{iat}:

$$\pounds\,(e^{iat}) = \int_{t=0}^{\infty} e^{-st}.\,e^{iat}.\,dt = \int_{t=0}^{\infty} e^{-(s-ia)t}\,dt = 1/(s-ia),$$

by Eq. (2.4); or, by definition of e^{iat} and rationalization of the RHS

$$\pounds\,(\cos at + i \sin at) = (s+ia)/(s-ia)(s+ia) = (s+ia)/(s^2+a^2).$$

Equating the real and imaginary parts we, thus, obtain

$$\pounds\,(\cos at) = s/(s^2+a^2), \qquad (2.5a)$$

$$\pounds\,(\sin at) = a/(s^2+a^2). \qquad (2.5b)$$

2.6. Hyperbolic functions:

$$£\,(\cosh at) = £\,(e^{at} + e^{-at})\,/\,2$$

$$= \{1\,/\,(s-a) + 1\,/\,(s+a)\}\,/\,2 = s\,/\,(s^2 - a^2), \qquad (2.6a)$$

by Eq. (2.4). Similarly, we obtain

$$£\,(\sinh at) = £\,(e^{at} - e^{-at})\,/\,2$$

$$= \{1\,/\,(s-a) - 1\,/\,(s+a)\}/\,2 = a\,/\,(s^2 - a^2), \qquad (2.6b)$$

again by Eq. (2.4).

Example 2.1. Find the Laplace transform of the functions:

(i) $\sin 2t.\ \sin 3t,$ and **(ii)** $\sin^3 2t.$

Solution. (i) Since $\sin 2t.\ \sin 3t = (\cos t - \cos 5t)\,/\,2,$

$$£\,(\sin 2t.\sin 3t) = \{£\,(\cos t) - £\,(\cos 5t)\}/\,2$$

$$= \{s\,/\,(s^2 + 1^2) - s\,/\,(s^2 + 5^2)\}\,/\,2, \qquad \text{by Eq. (2.5a)}$$

$$= 12\,s\,/\,(s^2 + 1^2)\,(s^2 + 5^2).$$

(ii) Since $\sin 6t = 3.\sin 2t - 4\sin^3 2t \Rightarrow \sin^3 2t = (3.\sin 2t - \sin 6t)/4,$

$$£\,(\sin^3 2t) = £\,(3.\sin 2t - \sin 6t)\,/\,4$$

$$= \{3.2/(s^2 + 2^2) - 6/(s^2 + 6^2)\}/4 = 48\,/\,(s^2 + 4)\,(s^2 + 36), \text{ by Eq. (2.5b).//}$$

§ 3. Some Properties of Laplace Transforms

3.1. Linear property: Given two functions $F(t)$ and $G\,(t)$ of a variable t the Laplace transform of their linear sum is given by

$$£\,\{a\,F\,(t) + b\,G\,(t)\} = a\,£\{F\,(t)\} + b\,£\{G\,(t)\}, \qquad (3.1)$$

where a, b are constants.

Proof. By the definition of Laplace transform, the LHS of Eq. (3.1) is

$$\int_{t=0}^{\infty} e^{-st}.\{a\,F(t) + b\,G(t)\}dt = a\int_{t=0}^{\infty} e^{-st}.F(t)\,dt + b\int_{t=0}^{\infty} e^{-st}.G(t)\,dt$$

$$= a\,£\,\{F(t)\} + b\,£\,\{G(t)\},$$

again by the definition of Laplace transform.

3.2. First shifting (or translation) property: Given a function $F(t)$ with its Laplace transform $£\,\{F(t)\} = f(s)$, we have

$$£\,\{e^{at}.\,F(t)\} = f(s-a). \tag{3.2}$$

Proof. By the definition of Laplace transform, the LHS of Eq. (3.2) is

$$\int_{t=0}^{\infty} e^{-st}.\,e^{at}.\,F(t)\,dt = \int_{t=0}^{\infty} e^{-(s-a)t}.\,F(t)\,dt$$

$$= \int_{t=0}^{\infty} e^{-ut}.\,F(t)\,dt = f(u) = f(s-a),$$

where we have put $s - a = u > 0$.

3.3. Second shifting property: Given a function $F(t)$ with its Laplace transform $£\,\{F(t)\} = f(s)$, the Laplace transform of the function $G(t)$ defined by

$$G(t) = F(t-a), \text{ when } t \geq a; \text{ and } G(t) = 0, \text{ when } t < a, \tag{3.3}$$

is given by
$$£\,\{G(t)\} = e^{-as}.\,f(s). \tag{3.4}$$

Proof. By definition

$$£\{G(t)\} = \int_{t=0}^{\infty} e^{-st}.G(t)\,dt = \int_{t=0}^{a-\varepsilon} 0.\,e^{-st}.\,dt + \int_{t=a}^{\infty} e^{-st}.F(t-a)\,dt.$$

Putting $t - a = u$ so that $dt = du$ and applying definite integrals' first property:

$$\int_{u=0}^{\infty} e^{-s(u+a)}.\,F(u)\,du = e^{-sa}\int_{u=0}^{\infty} e^{-su}.\,F(u)\,du$$

$$= e^{-sa} \int_{t=0}^{\infty} e^{-st} . F(t)\, dt = e^{-sa} . £ \{F(t)\},$$

which is the RHS of Eq. (3.4).

3.4. Change of scale property: Given a function $F(t)$ with its Laplace transform $£ \{F(t)\} = f(s)$, we have

$$£ \{F(a t)\} = (1/a). f(s/a). \qquad (3.5)$$

Proof. By definition, given by Eq. (1.5), and putting $a t = u$ so that $dt = (du)/a$, we get

$$£ \{F(at)\} = \int_{t=0}^{\infty} e^{-st} . F(at)\, dt = (1/a) \int_{u=0}^{\infty} e^{-(s/a)u} . F(u)\, du$$

$$= (1/a) \int_{t=0}^{\infty} e^{-(s/a)t} . F(t)\, dt = (1/a) f(s/a),$$

by a property of definite integrals.

§ 4. Examples Based on Above Properties of Laplace Transforms

Example 4.1. Find the Laplace transform of the function

$$F(t) = 3\, e^{-5t} + 4t^{6} - 4 \sin 4t + 2 \cos 5t.$$

Solution. By linear property,

$$£ F(t) = 3\, £ (e^{-5t}) + 4\, £ (t^{6}) - 4\, £ (\sin 4t) + 2\, £ (\cos 5t)$$

$$= 3/(s+5) + 4.(6!)/s^{7} - 4.\, 4/(s^{2} + 4^{2}) + 2s/(s^{2} + 5^{2}),$$

where the results of Section 2 have been used. //

Example 4.2. Find the Laplace transform of the functions:

(i) $\sin t . \cos t$, **(ii)** $(\sin t - \cos t)^{2}$, **(iii)** $\cos^{2} at$, **(iv)** $\cosh^{2} (2t)$.

Solution. (i) $F(t) = \sin t . \cos t = (\sin 2t)/2$. Therefore,

$$£ F(t) = £ (\sin 2t)/2 = 2/(s^{2} + 2^{2}).\, 2 = 1/(s^{2} + 4).$$

(ii) $F(t) = (\sin t - \cos t)^2 = \sin^2 t + \cos^2 t - 2 \sin t.\cos t = 1 - \sin 2t.$

Therefore, when $s > 0$,

$$£ F(t) = £(1) - £(\sin 2t) = 1/s - 2/(s^2 + 2^2) = 1/s - 2/(s^2 + 4).$$

(iii) $F(t) = \cos^2 at = (1 + \cos 2at)/2$. Therefore, for $s > 0$,

$$£ F(t) = \{£(1) + £(\cos 2at)\}/2 = \{1/s + s/(s^2 + 4a^2)\}/2.$$

(iv) $F(t) = \cosh^2(2t) = (1 + \cosh 4t)/2.$

Therefore, for $s > 4$,

$$£ F(t) = \{£(1) + £(\cosh 4t)\}/2 = \{1/s + s/(s^2 - 16)\}/2. //$$

Example 4.3. Find the Laplace transform of the function defined by

$F(t) = 0$ (when $0 < t < 2$), $2t$ (when $2 < t < 4$), 0 (when $4 < t$).

Solution. By definition of Laplace transform,

$$£ F(t) = \int_{t=0}^{\infty} e^{-st}.F(t)\,dt = \int_{t=0}^{2-\varepsilon} 0.e^{-st}.\,dt + \int_{t=2+\varepsilon}^{4-\varepsilon} 2t.e^{-st}.\,dt$$

$$+ \int_{t=4+\varepsilon}^{\infty} 0.e^{-st}.\,dt = (2/s).\{\left[-t.e^{-st}\right]_{2+\varepsilon}^{4-\varepsilon} + \int_{t=2+\varepsilon}^{4-\varepsilon} e^{-st}.\,dt\}$$

$$= (2/s).\{-4.e^{-4s} + 2.e^{-2s} - \left[e^{-st}/s\right]_{2+\varepsilon}^{4-\varepsilon}\}$$

$$= (2/s).\{-4\,e^{-4s} + 2\,e^{-2s} - e^{-4s}/s + e^{-2s}/s\}$$

$$= (2/s^2)\{(2s+1)\,e^{-2s} - (4s+1)\,e^{-4s}\}. //$$

Example 4.4. Find the Laplace transform of the function defined by

$F(t) = \cos(t - 2\pi/3)$, when $t > 2\pi/3$, and 0, when $t < 2\pi/3$.

Solution. By definition of Laplace transform,

$$£ F(t) = \int_{t=0}^{\infty} e^{-st}.F(t).\,dt$$

$$= \int_{t=0}^{2\pi/3-\varepsilon} 0.e^{-st}.\,dt + \int_{t=2\pi/3+\varepsilon}^{\infty} e^{-st}.\cos(t-2\pi/3).\,dt.$$

Putting $t - 2\pi/3 = u$ so that $dt = du$, above integral reduces to

$$\int_{u=\varepsilon}^{\infty} e^{-s(u+2\pi/3)}.\cos u.\,du = e^{-2\pi s/3}.\int_{u=\varepsilon}^{\infty} e^{-su}.\cos u.\,du$$

$$= e^{-2\pi s/3}.\,£\,(\cos u) = (e^{-2\pi s/3}.\,s)/(s^2+1), \qquad \text{for } s > 0.\ //$$

Example 4.5. Using the *first shifting property* find the Laplace transform of the function

$$F(t) = e^{-t}.(\cos 4t - 2.\sin 4t). \tag{4.1}$$

Solution. Applying the *first shifting property* and putting from Eq. (2.5), the Laplace transform of the function reads as

$$£\,F(t) = £\,(e^{-t}.\cos 4t) - 2.\,£\,(e^{-t}.\sin 4t) = f(s+1) - 2g(s+1),$$

where

$$f(s) = £\,(\cos 4t) = s/(s^2+16), \text{ and } g(s) = £\,(\sin 4t) = 4/(s^2+16).$$

Therefore,

$$£\,F(t) = (s+1)/\{(s+1)^2+16\} - 8/\{(s+1)^2+16\}$$

$$= (s-7)/(s^2+2s+17).\ //$$

§ 5. Laplace Transform of Derivative of a function

Let $F(t)$ be a continuous function of some variable t with its derivative $F'(t)$ section ally continuous but free from any discontinuities. An example of such a function $F(t)$ and its derivative $F'(t)$ is exhibited by Figure 5.1.

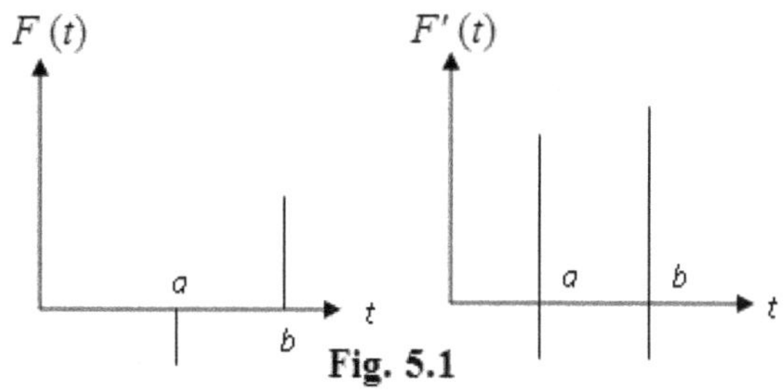

Fig. 5.1

Theorem 5.1. If $£\,F(t) = f(s)$ then

$$£\,F'(t) = s.f(s) - F(0). \tag{5.1}$$

Proof. By definition, $\pounds\, F'(t) = \int_{t=0}^{\infty} e^{-st}.F'(t)\, dt$. Integrating it by *method of parts* taking $F'(t)$ as the second function, the above integral simplifies to

$$\left[e^{-st}.F(t)\right]_0^{\infty} + s.\int_{t=0}^{\infty} e^{-st}.\, F(t)\, dt$$

$$= -F(0) + s.\, \pounds\, F(t) = -F(0) + s.f(s)$$

provided $s > 0$ and $\lim_{t\to\infty} \{e^{-st}.\, F(t)\} = 0$ is assumed. //

Corollary 5.1. Laplace transform of the second order derivative of a function $F(t)$ is given by

$$\pounds\, F''(t) = s^2.\pounds\, F(t) - s.F(0) - F'(0) = s^2.f(s) - s.F(0) - F'(0). \quad (5.2)$$

Proof. Assuming the continuity of the first order derivative $F'(t)$ and putting

$$F'(t) = G(t) \quad \Rightarrow \quad F''(t) = G'(t),$$

we evaluate the desired result by means of above theorem:

$$\pounds\, F''(t) = \pounds\, G'(t) = s.\, \pounds\, G(t) - G(0)$$

$$= s.\, \pounds\, F'(t) - F'(0) = s\, \{s.f(s) - F(0)\} - F'(0),$$

which is Eq. (5.2) only. //

Corollary 5.2. Laplace transform of derivative of a function $F(t)$ of any order, say n, is given by

$$\pounds\{F^{(n)}(t)\} = s^n\, \pounds\, F(t) - s^{n-1}\, F(0) - s^{n-2}\, F'(0)$$
$$- s^{n-3}\, F''(0) - \ldots - F^{(n-1)}(0), \quad (5.3)$$

where all the derivatives of the function up to $(n-1)^{\text{th}}$ order are assumed continuous and $F^{(n)}(t)$ is sectionally continuous. The limit

$$\lim_{t\to\infty} \{e^{-st}.\, F^{(n-1)}(t)\} = 0$$

is also assumed.

§ 6. Laplace Transform of Integrals

In this Section we derive the Laplace transform of a function which is expressed as an integral of another function.

Theorem 6.1. Laplace transform of a function $G(t)$:

$$G(t) \equiv \int_{u=0}^{t} F(u)\, du \tag{6.1}$$

is given by

$$£\, G(t) \equiv £\, \{\int_{u=0}^{t} F(u)\, du\} = \{£\, F(t)\} / s = \{f(s)\}/ s. \tag{6.2}$$

Proof. Differentiating Eq. (6.1) w.r.t. t, we get

$$G'(t) = F(t) \quad \Rightarrow \quad £\, G'(t) = £\, F(t). \tag{6.3}$$

We also note from Eq. (6.1):

$$G(0) \equiv \int_{u=0}^{0} F(u)\, du = 0. \tag{6.4}$$

Applying Eq. (5.1) to $G'(t)$:

$$£\, G'(t) = s.\, £\, G(t) - G(0),$$

or, by Eqs. (6.3) and (6.4),

$$£\, F(t) = s.£\, G(t) \quad \Rightarrow \quad £\, G(t) = (1/s)\, £\, F(t) = (1/s).\, f(s).\ //$$

Example 6.1. Assuming that £ (cos t) is known derive the Laplace transform of the function

$$F(t) = t.\sin t, \tag{6.5}$$

by applying the formula for the Laplace transform of a derivative.

Solution. Differentiating Eq. (6.5) twice w.r.t. t, we obtain

$$F'(t) = \sin t + t.\cos t, \qquad F''(t) = 2.\cos t - t.\sin t. \tag{6.6}$$

Noting $F(0) = 0$, $F'(0) = 0$, and applying the Eq. (5.2), we get

$$£\, F''(t) = s^2.\, £\, F(t),$$

or, for Eqs. (6.5) and (6.6),

$$£\,(2.\cos t - t.\sin t) = s^2.\,£\,(t.\sin t),$$

or,

$$2.\,£\,(\cos t) - £\,(t.\sin t) = s^2.\,£\,(t.\sin t)$$

$\Rightarrow$

$$£\,(t.\sin t) = 2\,(£\cos t)\,/\,(1+s^2) = 2s\,/\,(1+s^2)^2.\ // \qquad (6.7)$$

Example 6.2. Evaluate the function $F(t)$ if

$$£\,F(t) \equiv f(s) = 16\,/\,(s^2+4)^2, \qquad (6.8)$$

and

$$£\,(t.\sin 2t) = 4s\,/\,(s^2+4)^2. \qquad (6.9)$$

Solution. Applying Eq. (6.2) for

$$H(u) \equiv 4u.\sin 2u, \qquad (6.10)$$

we get

$$£\,\{\int_{u=0}^{t} (4u.\sin 2u)\,du\} = (1/s)\,£\,H(t) = (1/s)\,£\,(4t.\sin 2t)$$

$$= (4/s)\,£\,(t.\sin 2t) = 16\,/\,(s^2+4)^2, \qquad \text{by Eq. (6.9)}$$

$$= £\,F(t), \qquad \text{by Eq. (6.8)}$$

$\Rightarrow$

$$F(t) = \int_{u=0}^{t} (4u.\sin 2u)\,du = \int_{v=0}^{2t} (v.\sin v)\,dv = \left[-v.\cos v\right]_0^{2t}$$

$$+ \int_{v=0}^{2t} \cos v\,dv = -2t.\cos 2t + \left[\sin v\right]_0^{2t} = -2t.\cos 2t + \sin 2t,$$

where we have put $2u = v$ and, therefore, $2du = dv$. //

Example 6.3. Find Laplace transform of function $G(t) \equiv t.\,e^t.\sin t$.

Solution. Interpreting $G(t)$ as the product of e^t and $F(t) \equiv t.\sin t$, and applying the *first shifting property*, we get

$$£\,G(t) = £\,\{e^t.\,F(t)\} = f(s-1)$$

$$= 2\,(s-1)\,/\,\{1+(s-1)^2\}^2 = 2(s-1)\,/\,(s^2-2s+2)\}^2, \qquad \text{by Eq. (6.7). //}$$

§ 7. Laplace Transform of a Function $t^n . F(t)$

Theorem 7.1. If £ $\{F(t)\} = f(s)$, then

$$£\ \{t^n . F(t)\} = (-1)^n D^n f(s) \equiv (-1)^n d^n f(s)/ds^n, \ n = 1, 2, 3, \ldots \quad (7.1)$$

Proof. (i) For $n = 1$, the RHS of (7.1) becomes

$$-df(s)/ds = -(d/ds)\ £\ \{F(t)\} = -(d/ds)\left\{\int_{t=0}^{\infty} e^{-st} . F(t)\, dt\right\}.$$

Using the Leibniz rule for differentiation under the sign of integration the RHS of above relation simplifies to

$$\left\{\int_{t=0}^{\infty} t . e^{-st} . F(t)\, dt\right\} = \int_{t=0}^{\infty} e^{-st} . \{t . F(t)\}\, dt = £\ \{t . F(t)\}$$

by definition. The same is the LHS of (7.1) for $n = 1$. Thus, (7.1) holds good for $n = 1$:

$$£\ \{t . F(t)\} = -(d/ds)\ £\ \{F(t)\} = -df(s)/ds. \quad (7.1a)$$

(ii) Next, let the result given by Eq. (7.1) be true for any value, say $m < n$:

$$£\ \{t^m . F(t)\} = (-1)^m . D^m f(s). \quad (7.2)$$

Differentiating it further w.r.t. s, multiplying the result by -1 and interchanging the order of the result, we obtain

$$(-1)^{m+1} . D^{m+1} f(s) = -(d/ds)\ £\ \{t^m . F(t)\}$$

$$= -(d/ds)\left\{\int_{t=0}^{\infty} e^{-st} . t^m . F(t)\, dt\right\}, \qquad \text{by definition.}$$

Again applying the Leibniz rule for differentiation under the sign of integration the RHS of above relation further reduces to

$$\left\{\int_{t=0}^{\infty} t . e^{-st} . t^m . F(t)\, dt\right\}$$

$$= \int_{t=0}^{\infty} e^{-st} . \{t^{m+1} . F(t)\}\, dt = £\ \{t^{m+1} . F(t)\},$$

by definition. Thus, as a necessary consequence of Eq. (7.2), there also results

$$£\ \{t^{m+1}.\ F(t)\} = (-1)^{m+1}.D^{m+1} f(s). \tag{7.3}$$

Therefore, by the principle of mathematical induction, it is established that (7.1) holds for any positive integral value of n. //

§ 8. Laplace Transform of a Function $F(t)$ Divided by t

Theorem 8.1. If $£\ \{F(t)\} = f(s)$ and the following integral exists,

$$£\ [\{F(t)\}/t] = \int_{u=s}^{\infty} f(u)\, du. \tag{8.1}$$

Proof. Let

$$G(t) \equiv \{F(t)\}/t \quad \Rightarrow \quad t.\ G(t) \equiv F(t). \tag{8.2}$$

Forming the Laplace transforms of both sides of this relation, we get

$$£\ \{t.\ G(t)\} = £\ F(t) = f(s), \qquad \text{say.} \tag{8.3}$$

Applying (7.1a) for $G(t)$ and writing $£\ G(t) = g(s)$, say, above relation reduces to

$$£\ \{t.\ G(t)\} = -(d/ds)\ £\ \{G(t)\} = -(d/ds)\, g(s) = f(s), \qquad \text{by (8.3).}$$

This on integration w.r.t. s from s to ∞ yields

$$-\int_{s}^{\infty} dg(s) = [-g(s)]_{s}^{\infty} = g(s) = \int_{s}^{\infty} f(s)\, ds = \int_{u=s}^{\infty} f(u)\, du, \tag{8.4}$$

for

$$\lim_{s\to\infty} g(s) = \lim_{s\to\infty} £\ G(t) = \lim_{s\to\infty} \int_{t=0}^{\infty} e^{-st}.\ G(t)\, dt$$

$$= \int_{t=0}^{\infty} (\lim_{s\to\infty} e^{-st}).\ G(t)\, dt = \int_{t=0}^{\infty} 0.\ G(t)\, dt = 0,$$

as t is never negative on the given range of integration and s being positive. Thus, Eqs. (8.2) and (8.4) establish the equation (8.1). //

Example 8.1. Show that

$$£\ \{(\sin t)/t\} = \tan^{-1}(1/s). \tag{8.5}$$

Hence, deduce £ {(sin at) / t}.

Solution. (i) Firstly, from Eq. (2.5b), we find

$$£\,(\sin t) = 1/(s^2+1) \equiv f(s), \text{ say} \quad \Rightarrow \quad f(u) = 1/(u^2+1),$$

and next apply Theorem 8.1 for the function sin t:

$$£\,\{(\sin t)/t\} = \int_{u=s}^{\infty} f(u)\,du = \int_{u=s}^{\infty} \{1/(u^2+1)\}.\,du. \qquad (8.6)$$

Putting $u = 1/v$ so that $du = -(1/v^2)\,dv$, above integral reduces to

$$-\int_{v=1/s}^{0} \{1/(v^2+1)\}.\,dv = \int_{v=0}^{1/s} \{1/(v^2+1)\}\,dv$$

$$= \left[\tan^{-1} v\right]_0^{1/s} = \tan^{-1}(1/s).$$

Alternately, the integral in (8.6) can also be evaluated as

$$\left[\tan^{-1} u\right]_{u=s}^{\infty} = \tan^{-1}\infty - \tan^{-1} s = \pi/2 - \tan^{-1} s = \alpha \text{ (say)}$$

$\Rightarrow$

$$\pi/2 - \alpha = \tan^{-1} s \quad \Rightarrow \quad \tan(\pi/2 - \alpha) = \cot\alpha = s$$

$\Rightarrow$

$$\tan\alpha = 1/s \qquad \Rightarrow \alpha = \tan^{-1}(1/s).$$

(ii) Applying the *change of scale property* given by Eq. (3.5) to the function $F(t) \equiv (\sin t)/t$, and using Eq. (8.5), we deduce

$$£\,F(at) = £\,\{(\sin at)/at\} = (1/a)\tan^{-1}\{1/(s/a)\} = (1/a).\tan^{-1}(a/s).$$

Multiplying on both sides by a, we get

$$£\,\{(\sin at)/t\} = \tan^{-1}(a/s).\ // \qquad (8.7)$$

Example 8.2. Find the Laplace transform of the function $(1 - e^t)/t$.

Solution. By Eqs. (2.1) and (2.4), we have

$$£\,(1 - e^t) = 1/s - 1/(s-1) \equiv f(s), \text{ say } \Rightarrow f(u) = 1/u - 1/(u-1).$$

Therefore, Theorem 8.1 for the function $F(t) \equiv 1 - e^t$ yields the following relation

$$£\,\{(1 - e^t)/t\} = \int_{u=s}^{\infty} f(u)\,du = \int_{u=s}^{\infty} \{1/u - 1/(u-1)\}.du$$

$$= \left[\log_e u - \log_e (u-1)\right]_s^{\infty} = \left[\log_e \{u/(u-1)\}\right]_s^{\infty} = -\ln\{s/(s-1)\}$$

$$= \log_e \{(s-1)/s,$$

as

$$\lim_{u\to\infty} \log_e \{u/(u-1)\} = \lim_{u\to\infty} \log_e \{1/(1 - 1/u)\} = \log_e 1 = 0.//$$

Example 8.3. Given $£\,(2\sqrt{(t/\pi)}\} = 1/s^{3/2}$, show that

$$£\,\{1/\sqrt{(\pi t)}\} = 1/\sqrt{s}.$$

Solution. Let $F(t) = 2\sqrt{(t/\pi)} \Rightarrow \{F(t)\}/t = 2/\sqrt{(\pi t)}$. Thus, by Theo. (8.1),

$$£\,[\{F(t)\}/t] = £\,\{2/\sqrt{(\pi t)}\} = \int_{u=s}^{\infty} f(u)\,du = \int_{u=s}^{\infty} £\,F(t)\,du$$

$$= \int_{u=s}^{\infty} \{1/u^{3/2}\}\,du = -2\left[u^{-1/2}\right]_s^{\infty} = 2/\sqrt{s}.$$

Dividing both sides by 2, we get the desired result. //

Example 8.4. Evaluate the integral $\int_{t=0}^{\infty} t^3.e^{-t}.\sin t\,dt$ by techniques of Laplace transform.

Solution. Applying (7.1) for $t^3.\sin t$ we have

$$£\,(t^3.\sin t) = (-1)^3.D^3\,£\,(\sin t) = -D^3\{1/(1+s^2), \qquad \text{by Eq. (2.5b)}$$

i.e.

$$\int_{t=0}^{\infty} e^{-st}.(t^3.\sin t)\,dt = D^2\{2s/(1+s^2)^2\}$$

$$= D\{2/(1+s^2)^2 - 8s^2/(1+s^2)^3\}$$

$$= -24s/(1+s^2)^3 + 48s^3/(1+s^2)^4 = 24s\,(s^2-1)/(1+s^2)^4.$$

Taking $s = 1$ we, thus, obtain $\int_{t=0}^{\infty} t^3.e^{-t}.\sin t\, dt = 0.$ //

Example 8.5. Evaluate the integral $\int_{t=0}^{\infty} \{(\sin t) / t\}\, dt$ by Laplace transform method.

Solution. Applying definition of Laplace transform on the LHS and evaluating the integral on the RHS, Eq. (8.6) reduces to

$$\int_{t=0}^{\infty} \{ e^{-st}. (\sin t) / t \}\, dt = \left[\tan^{-1} u\right]_s^{\infty} = \pi/2 - \tan^{-1} s.$$

Taking limits as $s \to 0$, we derive the result:

$$\int_{t=0}^{\infty} \{(\sin t) / t\}\, dt = \pi / 2. \text{ //} \qquad (8.8)$$

§ 9. Two Important Theorems

Theorem 9.1 (*Initial value theorem*). Let $F(t)$ be a continuous function for all values of $t \geq 0$ and be of exponential order as $t \to \infty$. Further, if $F'(t)$ is sectionally continuous and of some exponential order then

$$\lim_{t \to 0} F(t) = F(0) = \lim_{s \to \infty} s.f(s) = \lim_{s \to \infty} \{s. \pounds\, F(t)\}. \quad (9.1)$$

Proof. Laplace transform of the derivative $F'(t)$ is given by Eq. (5.1); which, as per definition of the Laplace transform, reads as

$$\int_{t=0}^{\infty} e^{-st}. F'(t)\, dt = s.f(s) - F(0). \qquad (9.2)$$

Taking limit as $s \to \infty$ above relation yields

$$\int_{t=0}^{\infty} \{\lim_{s \to \infty} e^{-st}\}\, F'(t)\, dt = \lim_{s \to \infty} \{s.f(s) - F(0)\}.$$

Both s and t being positive the LHS integral vanishes for above choice of $F'(t)$. Hence, the RHS establishes the desired result. //

Theorem 9.2 (*Final value theorem*). Under the conditions as stated in the previous theorem, there holds

$$\lim_{t \to \infty} F(t) = \lim_{s \to 0} s f(s) = \lim_{s \to 0} \{s \pounds F(t)\}. \qquad (9.3)$$

Proof. Proceeding as in Theorem 9.1, and taking limit as $s \to 0$, Eq. (9.2) yields

$$\lim_{s \to 0} \{ s. f(s) - F(0) \} = \lim_{s \to 0} \{ \int_{t=0}^{\infty} e^{-st}. F'(t)\, dt \}$$

$$= \int_{t=0}^{\infty} F'(t)\, dt = [F(t)]_0^{\infty} = \lim_{t \to \infty} F(t) - F(0),$$

giving the desired result. //

§ 10. Laplace Transforms of Some Special Functions

10.1. The error function: The *error function* abbreviated as $E(t)$ or $erf(t)$ is defined by

$$E(t) \equiv erf(t) \equiv (2/\sqrt{\pi}) \int_{u=0}^{t} \exp(-u^2)\, du. \qquad (10.1)$$

Its complement $erf_c(t)$ is defined by

$$erf_c(t) \equiv 1 - erf(t) = 1 - (2/\sqrt{\pi}) \int_{u=0}^{t} \exp(-u^2)\, du$$

$$= 1 - (2/\sqrt{\pi}) \{ \int_{u=0}^{\infty} \exp(-u^2)\, du + \int_{u=\infty}^{t} \exp(-u^2)\, du \}. \qquad (10.2)$$

Putting $u^2 = x$ so that $u = \sqrt{x}$ and $du = (1/2\sqrt{x})\, dx$, the first integral in Eq. (10.2) reduces to:

$$\int_{u=0}^{\infty} \exp(-u^2)\, du = (1/2) \int_{x=0}^{\infty} e^{-x}. x^{-1/2}\, dx$$

$$= (1/2) \int_{x=0}^{\infty} x^{(1/2-1)}. e^{-x}. dx = (1/2)\, \Gamma(1/2) = (\sqrt{\pi})/2. \qquad (10.3)$$

Hence, equation (10.2) simplifies to as:

$$erf_c(t) = (2/\sqrt{\pi}) \int_{u=t}^{\infty} \exp(-u^2)\, du. \qquad (10.4)$$

Theorem 10.1. $\pounds\ erf(\sqrt{t}) = 1/s\sqrt{(s+1)}$. (10.5)

Proof. By Eq. (10.1),

$$erf\ (\sqrt{t}) = (2/\sqrt{\pi})\int_{u=0}^{\sqrt{t}} \exp(-u^2)\,du$$

$$= (2/\sqrt{\pi})\int_{u=0}^{\sqrt{t}} \{1 - u^2 + u^4/2! - u^6/3! + \ldots\}\,du$$

$$= (2/\sqrt{\pi})\,\left[u - u^3/3 + u^5/5.2! - u^7/7.3! + \ldots\right]_0^{\sqrt{t}}$$

$$= (2/\sqrt{\pi})\,\{t^{1/2} - t^{3/2}/3 + t^{5/2}/5.2! - t^{7/2}/7.3! \ldots\}.$$

Therefore, by Eq. (2.3)

$$£\ erf\ (\sqrt{t}) = (2/\sqrt{\pi})\,\{\Gamma(3/2)/s^{3/2} - \Gamma(5/2)/3s^{5/2}$$

$$+ \Gamma(7/2)/5.2!\,s^{7/2} - \Gamma(9/2)/7.3!\,s^{9/2} + \ldots\}$$

$$= 1/s^{3/2} - 1/2s^{5/2} + (1.3)/2.4\,s^{7/2} - (1.3.5)/2.4.6\,s^{9/2} + \ldots$$

$$= (1/s^{3/2})\,\{1 + (-1/2)(1/s) + (-1/2)(-3/2)(1/s)^2/2!$$

$$+ (-1/2)(-3/2)(-5/2)(1/s)^3/3! + \ldots\}$$

$$= (1/s^{3/2})(1 + 1/s)^{-1/2} = (1/s^{3/2})\,\{s/(s+1)\}^{1/2} = 1/s\sqrt{(s+1)}.\ //$$

10.2. Laplace transform of *sine* integral: *Sine* integral is defined by

$$S_i(t) \equiv \int_{u=0}^{t} \{(\sin u)/u\}\,du. \qquad (10.6)$$

Theorem 10.2. $£\ S_i(t) = (1/s)\tan^{-1}(1/s).$ (10.7)

Proof. Setting $F(t) = S_i(t)$ and differentiating it w.r.t. t, we get

$$F'(t) = (\sin t)/t \quad \Rightarrow \quad t.F'(t) = \sin t. \qquad (10.8)$$

Forming its Laplace transform and applying (7.1a), we obtain

$$£\ \{t.F'(t)\} = -(d/ds)\,\{£\ F'(t)\} = £\ (\sin t),$$

or, for Eqs. (2.5b) and (5.1)

$$(d/ds)\,\{s\,f(s) - F(0)\} = (d/ds)\,\{s\,f(s)\} = -1/(1+s^2), \qquad (10.9)$$

as

$$F(0) = S_i(0) = \int_{u=0}^{0} \{(\sin u)/u\}\, du = 0. \qquad (10.10)$$

Integrating Eq. (10.9) w.r.t. s, we evaluate

$$s f(s) = \cot^{-1} s + a \qquad (10.11)$$

where a is some arbitrary constant of integration. Taking limit of Eq. (10.11) as $s \to \infty$, applying the *Initial value theorem* and putting from Eq. (10.10), we evaluate a:

$$\lim_{s\to\infty} s f(s) = \lim_{t\to 0} F(t) = F(0) = 0$$

$$= \cot^{-1} \infty + a = 0 + a \quad \Rightarrow \quad a = 0.$$

Hence, Eq. (10.11) determines

$$f(s) \equiv £\, F(t)\} = £\ S_i\ (t) = (1/s).\cot^{-1} s = (1/s)\tan^{-1}(1/s).//$$

10.3. Laplace transform of *cosine* integral: *Cosine* integral is defined by

$$C_i\ (t) \equiv \int_{u=t}^{\infty} \{(\cos u)/u\}\, du. \qquad (10.12)$$

Theorem 10.3. $£\, C_i(t) = \{\ln(1+s^2)\}/2s. \qquad (10.13)$

Proof. Setting

$$F(t) = C_i(t) = -\int_{u=\infty}^{t} \{(\cos u)/u\}\, du,$$

and differentiating it w.r.t. t, we get

$$F'(t) = (-\cos t)/t \quad \Rightarrow \quad t.F'(t) = -\cos t, \qquad (10.14)$$

as $\lim_{u\to\infty} (\cos u)/u = 0$. Taking the Laplace transform of Eq. (10.14) and putting from Eq. (7.1), we obtain

$$£\,\{t.F'(t)\} = -(d/ds)\,\{£\, F'(t)\} = £\,(-\cos t),$$

or, for Eqs. (2.5a) and (5.1)

$$(d/ds)\{s\,f(s) - F(0)\} = (d/ds)\{s\,f(s)\} = s/(1+s^2), \quad (10.15)$$

as $F(0)$ is constant making $(d/ds)\,F(0) = 0$. Integrating (10.15) w.r.t. s, we evaluate

$$s\,f(s) = (1/2)\ln(1+s^2) + b, \quad (10.16)$$

where b is some arbitrary constant of integration. Taking limit of Eq. (10.16) as $s \to 0$, applying the *Final value theorem*, we evaluate b:

$$\lim_{t\to\infty} F(t) = \lim_{t\to\infty} \int_{u=t}^{\infty} \{(\cos u)/u\}\,du$$

$$= \int_{u=\infty}^{\infty} \{(\cos u)/u\}\,du = 0 = b.$$

Hence, Eq. (10.16) determines

$$f(s) \equiv £\,F(t) = £\;C_i\;(t) = (1/2s)\ln(1+s^2). \;//$$

10.4. Laplace transform of exponential integral: The exponential integral, denoted by $E_i\,(t)$, is defined by

$$F(t) = E_i\,(t) \equiv \int_{u=t}^{\infty} (e^{-u}/u)\,du. \quad (10.17)$$

Theorem 10.4. $£\,E_i\,(t) = \{\ln(s+1)\}/s. \quad (10.18)$

Proof. Rewriting Eq. (10.17) as $F(t) = -\int_{u=\infty}^{t} (e^{-u}/u)\,du$ and differentiating it w.r.t. t, we get

$$F'(t) = -e^{-t}/t \quad \Rightarrow \quad t.\,F'(t) = -e^{-t}. \quad (10.19)$$

Taking the Laplace transform of (10.19) and putting from (7.1), we obtain

$$£\,\{t.\,F'(t)\} = -(d/ds)\,\{£\,F'(t)\} = £\,(-e^{-t}),$$

or, for Eqs. (2.4) and (5.1)

$$(d/ds)\,\{s\,f(s) - F(0)\} = (d/ds)\,\{s\,f(s)\} = 1/(s+1), \quad (10.20)$$

as $F(0)$ is constant making $(d/ds)\,F(0) = 0$. Integrating Eq. (10.20) w.r.t. s, we evaluate

$$s\,f(s) = \ln(s+1) + c, \tag{10.21}$$

where c is some arbitrary constant of integration. Taking limit of Eq. (10.21) as $s \to 0$, applying the *Final value theorem*, we evaluate c:

$$\lim_{t\to\infty} F(t) = \lim_{t\to\infty} \int_{u=t}^{\infty} (e^{-u}/u)\, du = \int_{u=\infty}^{\infty} (e^{-u}/u)\, du = 0 = c.$$

Hence, Eq. (10.21) determines

$$f(s) \equiv \pounds\, F(t)\} = \pounds\, E_i(t) = \{\log_e(s+1)\}/s. \;//$$

10.5. Laplace transform of Bessel's function: The Bessel's function of order n, denoted by $J_n(t)$, is defined by

$$J_n(t) = \{t^n/2^n.\Gamma(n+1)\}\{1 - t^2/2.(2n+2)$$

$$+ t^4/2.4.(2n+2).(2n+4) - \ldots \infty\}. \tag{10.22}$$

As a special case, for $n = 0$, the Bessel's function of order zero is

$$J_0(t) = 1 - t^2/2^2 + t^4/2^2.4^2 - t^6/2^2.4^2.6^2 + \ldots \infty$$

$$= \sum_{r=0}^{\infty} (-1)^r.\, t^{2r}/2^{2r}.(r!)^2. \tag{10.23}$$

Theorem 10.5. $\pounds\, J_0(t) = 1/\sqrt{(1+s^2)}$. (10.24)

Proof. Forming Laplace transform of Eq. (10.23) and applying Eq. (2.3), we get

$$\pounds\, J_0(t) = \pounds(1) - \pounds(t^2)/2^2 + \pounds(t^4)/2^2.4^2 - \pounds(t^6)/2^2.4^2.6^2 + \ldots \infty$$

$$= 1/s - \Gamma(3)/2^2 s^3 + \Gamma(5)/2^2.4^2.s^5 - \Gamma(7)/2^2.4^2.6^2.s^7 + \ldots \infty$$

$$= (1/s)\{1 - 1/2s^2 + (1.3)/2.4.s^4 - (1.3.5)/2.4.6.s^6 + \ldots\infty\}$$

$$= (1/s)(1 + 1/s^2)^{-1/2} = 1/\sqrt{(1+s^2)}. \;//$$

10.6. Some deductions from £ J_0 (t):

(i) Applying the change of scale property given by Eq. (3.5), the Eq. (10.24) gives

$$£\, J_0\,(at) \;=\; 1\,/\,a\,\surd\,(1+s^2/a^2) \;=\; 1\,/\,\surd\,(s^2+a^2). \qquad (10.25)$$

(ii) Applying Eq. (7.1a), we have

$$£\,\{t.\,J_0\,(at)\} = -\,(d/ds)\,£\,J_0\,(at) = -\,(d/ds)\,(s^2+a^2)^{-1/2} = s\,/\,(s^2+a^2)^{3/2}. \qquad (10.26)$$

(iii) Applying the *first shifting property*, the result given by Eq. (10.25) yields

$$£\,\{e^{-at}.\,J_0\,(at)\} = 1\,/\,\surd\,\{\,(s+a)^2+a^2\,\}. \qquad (10.27)$$

(iv) For definition of Laplace transform, Eq. (10.24) can be rewritten as

$$£\,J_0\,(t) \;=\; \int_{t=0}^{\infty} e^{-st}.\,J_0\,(t)\,dt \;=\; 1\,/\,\surd\,(1+s^2).$$

Taking its limit as $s \to 0$, we derive

$$\int_{t=0}^{\infty} J_0\,(t)\,dt \;=\; 1. \qquad (10.28)$$

Example 10.1. Evaluate £ (t. cos at).

Solution. Setting $F\,(t) =$ cos at, we have

$$£\;F\,(t) \equiv f\,(s) \;=\; £\,(\cos at) \;=\; s\,/\,(s^2+a^2). \qquad (10.29)$$

Applying Eq. (7.1a), we derive

$$£\,(t.\,\cos at) = -\,(d/ds)\,£\,(\cos at) = -\,(d/ds)\{s\,/\,(s^2+a^2)\}, \text{ by Eq. (10.29)}$$

$$= \{-\,(s^2+a^2)+2s^2)\}/\,(s^2+a^2)^2 = (s^2-a^2)\,/\,(s^2+a^2)^2.\;// \qquad (10.30)$$

Example 10.2. Evaluate the integral $\int_{t=0}^{\infty} t.\,e^{-3t}.\,\cos 4t.\,dt$.

Solution. By (10.30) and the definition of Laplace transform, we have

$$\pounds\,(t.\cos at) \equiv \int_{t=0}^{\infty} e^{-st}.\,t.\cos at.\,dt = (s^2 - a^2)/(s^2 + a^2)^2. \quad (10.31)$$

For $s = 3$ and $a = 4$, above relation yields the desired result: $-7/625$. //

Example 10.3. Show that

$$\int_{t=0}^{\infty} \{F(t)/t\}\,dt = \int_{u=0}^{\infty} f(u)\,du, \qquad (10.32)$$

and hence deduce that

$$\int_{t=0}^{\infty} \{(\sin t)/t\}\,dt = \pi/2. \qquad (10.33)$$

Solution. (i) Rewriting the equation (8.1), by definition,

$$\pounds\,[\{F(t)\}/t] \equiv \int_{t=0}^{\infty} \{e^{-st}.\,F(t)/t\}\,dt = \int_{u=s}^{\infty} f(u)\,du.$$

Taking its limit as $s \to 0$, we derive the desired result.

(ii) Taking $F(t) = \sin t$ so that

$$f(s) \equiv \pounds\,F(t) = \pounds\,(\sin t) = 1/(s^2 + 1), \text{ by Eq. (2.5b)}$$

or, equivalently

$$f(u) = 1/(u^2 + 1),$$

we derive from (10.32),

$$\int_{t=0}^{\infty} \{(\sin t)/t\}\,dt = \int_{u=0}^{\infty} f(u)\,du$$

$$= \int_{u=0}^{\infty} \{1/(u^2 + 1)\}\,du = \left[\tan^{-1} u\right]_0^{\infty} = \pi/2. \text{ //}$$

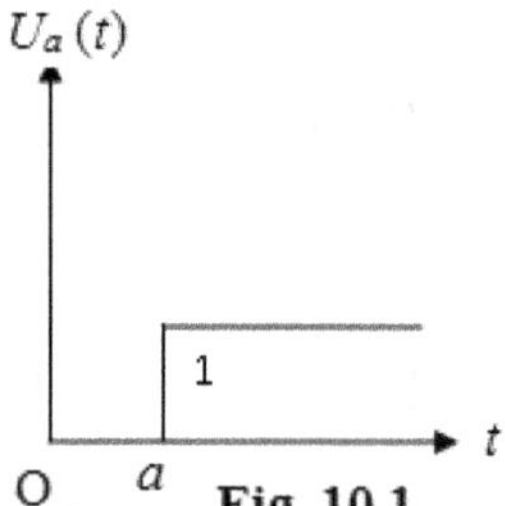

Fig. 10.1

10.7. Unit step function: The *unit step function*, also called the *Heaviside's unit function* and denoted by $U_a(t)$ or $U(t - a)$, is defined by

$$U_a(t) = 0 \text{ (when } t < a), 1 \text{ (when } t \geq a). \qquad (10.34)$$

Theorem 10.6. $$\pounds\,U_a(t) = e^{-as}/s. \qquad (10.35)$$

Proof. By definition,

$$£\, U_a(t) = \int_{t=0}^{\infty} e^{-st}.\, U_a(t).\, dt$$

$$= \int_{t=0}^{a} e^{-st}.\, 0.\, dt + \int_{t=a}^{\infty} e^{-st}.\, dt = -\left[e^{-st}/s\right]_a^{\infty} = e^{-as}/s. //$$

§ 11. Problem Set

11.1. Find the Laplace transforms of the following functions:

(i) $1 + 2\sqrt{t} + 3/\sqrt{t}$, **(ii)** $e^{2t} + 4t^3 - 2.\sin 3t + 3.\cos 3t$, **(iii)** $\sin^2 3t$,

(iv) $\cos^3 2t$, **(v)** $\cos(at + b)$, **(vi)** $\sin 2t.\cos 3t$,

(vii) $\cosh at - \cos at$, **(viii)** $t - \sinh 2t$.

11.2. Find the Laplace transforms of the functions:

(i) $t^3.\, e^{-3t}$, **(ii)** $(t + 2)^2.e^{t}$, **(iii)** $e^{-2t}.\sin 4t$,

(iv) $e^{-at}.\sinh bt$, **(v)** $e^{-t}.\sin^2 t$, **(vi)** $e^{-2t}.\sin 4t$,

(vii) $\cosh at.\sin at$, **(viii)** $e^{-3t}.(2.\cos 5t - 3.\sin 5t)$.

11.3. Find the Laplace transforms of the functions:

(i) $F(t) = e^t$ (when $0 < t < 1$), 0 (when $1 < t$),

(ii) $F(t) = \sin t$ (when $0 < t < \pi$), 0 (when $\pi < t$),

(iii) $F(t) = t^2$ (when $0 < t < 2$), $t - 1$ (when $2 < t < 3$), 7 (when $3 < t$),

(iv) $F(t) = |t - 1| + |t + 1|$, $0 \le t$,

(v) $F(t) = t/a$ (when $0 < t < a$), 1 (when $a < t$),

11.4. Show that $£\,(t.\sin at) = 2as/(s^2 + a^2)^2$.

11.5. Find the Laplace transforms of the functions:

(i) $t.\sin^2 t$, **(ii)** $t.\sin 3t.\cos 2t$, **(iii)** $t.\sinh at$, **(iv)** $t^2.\cos at$,

(v) $t. e^{-t}. \cosh t$, **(vi)** $t. e^{-2t}. \sin 2t$, **(vii)** $t^2. e^{-2t}. \cos t$,

(viii) $(e^{-at} - e^{-bt})/t$, **(ix)** $(e^{at} - \cos 6t)/t$, **(x)** $(e - t.\sin t)/t$,

(xi) $(1 - \cos 2t)/t$, **(xii)** $(1 - \cos t)/t^2$, **(xiii)** $(\cos 2t - \cos 3t)/t$.

11.6. Find the Laplace transforms of the following integrals:

(i) $\int_{t=0}^{t} e^{-t}. \cos t .dt$, **(ii)** $\int_{t=0}^{t} \{e^t. \sin t) / t\}. dt$,

(iii) $\int_{u=0}^{t} \{(1 - e^{-u}) / u\} du$.

11.7. Evaluate the following integrals by techniques of Laplace transform:

(i) $\int_{t=0}^{\infty} t. e^{-3t}. \sin t. dt$, **(ii)** $\int_{t=0}^{\infty} t. e^{-2t}. \cos t. dt$,

(iii) $\int_{t=0}^{\infty} \{(e^{-t} - e^{-3t}) / t\}. dt$, **(iv)** $\int_{t=0}^{\infty} \{(e^{-t}.\sin^2 t) / t\} dt$,

(v) $\int_{t=0}^{\infty} (\sin^2 t) / t^2\} dt$, **(vi)** $\int_{t=0}^{\infty} \{(\cos 6t - \cos 4t) / t\} dt$.

§ 12. Important formulae

Sr. No.	Function $F(t)$	Laplace transform £ $F(t)$ $= f(s)$
1.	1	$1 / s$
2.	t	$1 / s^2$
3.	t^n	$\Gamma(n+1)/ s^{n+1} = (n!)/ s^{n+1}$, if $n = 0, 1, 2, 3, \ldots$
4.	$t^{-1/2}$	$\Gamma(1/2) / \sqrt{s} = \sqrt{(\pi / s)}$
5.	$t^{1/2}$	$\Gamma(3/2) / s^{3/2} = \sqrt{\pi} / 2s^{3/2}$
6.	e^{at}	$1 / (s - a), \; s > a$
7.	$\cos at$	$s / (s^2 + a^2)$
8.	$\sin at$	$a / (s^2 + a^2)$
9.	$\cosh at$	$s / (s^2 - a^2)$

10.	sinh at	$a / (s^2 - a^2)$
11.	$a. F(t) + b. G(t)$	$a.$ £ $F(t) + b.$ £ $G(t)$
12.	$e^{at}. F(t)$	$f(s-a)$
13.	$e^{at}. \cos bt$	$(s-a) / \{(s-a)^2 + b^2\}$
14.	$e^{at}. \sin bt$	$b / \{(s-a)^2 + b^2\}$
15.	$F(at)$	$\{f(s/a)\} / a$
16.	$d F(t) / dt$	$s f(s) - F(0)$
17.	$d^2 (t) / dt^2$	$s^2 f(s) - s. F(0) - F'(0)$
18.	$\int_{u=0}^{t} F(u)\, du$	{£ $F(t)\} / s = \{f(s)\} / s$
19.	$t^n. F(t)$	$(-1)^n. D^n f(s), n = 1, 2, 3, \ldots$
20.	$t. F(t)$	$-df(s) / ds$
21.	$t. e^{at}$	$1/ (s-a)^2$
22.	$t. \cos at$	$(s^2 - a^2) / (s^2 + a^2)^2$
23.	$t. \sin at$	$2as / (s^2 + a^2)^2$
24.	$F(t) / t$	$\int_{u=s}^{\infty} f(u)\, du$
25.	$E(\sqrt{t}) \equiv erf(\sqrt{t}) = (2 / \sqrt{\pi}) \int_{u=0}^{\sqrt{t}} \exp(-u^2) du$	$1 / s \sqrt{(s+1)}$
26.	$S_i(t) \equiv \int_{u=0}^{t} \{(\sin u) / u\}\, du$	$\{\tan^{-1}(1/s)\} / s$
27.	$C_i(t) \equiv \int_{u=0}^{t} \{(\cos u) / u\} du$	$\{\ln(1 + s^2)\} / 2s$
28.	$E_i(t) \equiv \int_{u=t}^{\infty} (e^{-u}/u)\, du$	$\{\ln(1 + s)\} / s$
29.	(Bessel's function of order 0) $J_0(t)$	$1 / \sqrt{(1 + s^2)}$
30.	$J_0(at)$	$1 / \sqrt{(s^2 + a^2)}$
31.	$e^{-at}. J_0(at)$	$1 / \sqrt{\{(s+a)^2 + a^2\}}$
32.	Heaviside's unit function $U_a(t)$	e^{-as} / s

CHAPTER 7

INVERSE LAPLACE TRANSFORM

§ 1. Introduction

Definition 1.1. If the Laplace transform of a function $F(t)$ is $f(s)$:

$$£\{F(t)\} = f(s), \tag{1.1}$$

then $F(t)$ is called the *inverse Laplace transform* of $f(s)$ and we write

$$F(t) = £^{-1}\{f(s)\}, \tag{1.2}$$

where $£^{-1}$ denotes the inverse Laplace transform operator.

Definition 1.2. Given a function $N(t)$ of a variable t satisfying

$$\int_0^t N(t)\,dt = 0 \tag{1.3}$$

for all $t > 0$, is called a *null function.*

It may be noted that the Laplace transform of a null function is zero:

$$£\{N(t)\} \equiv \int_{t=0}^{\infty} e^{-st}.N(t)\,dt = 0. \tag{1.4}$$

Carrying out the integration by the method of parts taking $N(t)$ as the second function above integral reduces to

$$£\{N(t)\} = \left[e^{-st}.\int N(t)dt\right]_{t=0}^{\infty} + s\int_{t=0}^{\infty} \{e^{-st}.\int N(t)\,dt\}\,dt = 0,$$

by Eq. (1.3). Accordingly,

$$£\{F(t) + N(t)\} = £\{F(t)\} + £\{N(t)\} = £\{F(t)\} = f(s)$$

$\Rightarrow$

$$£^{-1}\{f(s)\} = F(t) = F(t) + N(t). \tag{1.5}$$

Theorem 1.1 (Lerch's theorem). If $F_1(t)$ and $F_2(t)$ are two functions having the same Laplace transform $f(s)$, then

$$F_1(t) - F_2(t) \equiv N(t) \tag{1.6}$$

is a null function for all $t > 0$.

Operating Eq. (1.6) by £ and noting Eq. (1.4), we conclude £ $F_1(t) =$ £ $F_2(t) = f(s)$, ensuring the uniqueness of the inverse Laplace transform except for the addition of a null function. //

§ 2. Properties of Inverse Laplace Transform

Theorem 2.1 (Linear property). If $f_1(s)$ and $f_2(s)$ are Laplace transforms of two functions $F_1(t)$ and $F_2(t)$ respectively, then

$$£^{-1}\{c_1.f_1(s) + c_2.f_2(s)\} = c_1.£^{-1}\{f_1(s)\} + c_2.£^{-1}\{f_2(s)\}$$

$$= c_1.F_1(t) + c_2.F_2(t), \tag{2.1}$$

where c_1, c_2 are any arbitrary constants.

Proof. By the linear property exhibited by Eq. (6.3.1) of Laplace transform of functions, we have

$$£\{c_1.F_1(t) + c_2.F_2(t)\} = c_1.£\{F_1(t)\} + c_2.£\{F_2(t)\}$$

$$= c_1.f_1(s) + c_2.f_2(s)$$

$\Rightarrow$

$$c_1.F_1(t) + c_2.F_2(t) = £^{-1}\{c_1.f_1(s) + c_2.f_2(s)\},$$

or, equivalently

$$c_1.£^{-1}\{f_1(s)\} + c_2.£^{-1}\{f_2(s)\} = £^{-1}\{c_1.f_1(s) + c_2.f_2(s)\}$$

which is Eq. (2.1). //

Theorem 2.2 (*First shifting property*). If $£^{-1}\{f(s)\} = F(t)$, then

$$£^{-1}\{f(s-a)\} = e^{at}.F(t) = e^{at}.£^{-1}\{f(s)\}. \tag{2.2}$$

Proof. Noting the definition of £ $\{F(t)\}$ given by Eq. (6.1.5), we have

$$f(s-a) = \int_{t=0}^{\infty} e^{-(s-a)t}.F(t)\,dt$$

$$= \int_{t=0}^{\infty} e^{-st}.\{e^{at}.F(t)\}\,dt = £\{e^{at}.F(t)\},$$

by definition of Laplace transform. This implies Eq. (2.2). //

Corollary 2.1. Dividing Eq. (2.2) by e^{at}, there also follows

$$£^{-1}\{f(s)\} = F(t) = e^{-at}.£^{-1}\{f(s-a)\}, \qquad (2.3)$$

or, on replacement of the parameter a by $-a$

$$£^{-1}\{f(s)\} = F(t) = e^{at}.£^{-1}\{f(s+a)\}. \qquad (2.4)$$

Theorem 2.3 (*Second shifting property*). If $£^{-1}\{f(s)\} = F(t)$, then

$$£^{-1}\{e^{-as}.f(s)\} = G(t) \equiv F(t-a) \text{ when } t \geq a, \; 0 \text{ when } t < a. \qquad (2.5)$$

Proof. Multiplying Eq. (6.1.5) by e^{-as}, we derive

$$e^{-as}.f(s) = \int_{t=0}^{\infty} e^{-s(t+a)}.F(t)\,dt = \int_{x=a}^{\infty} e^{-sx}.F(x-a).\,dx, \quad (2.6)$$

where we have put $t + a = x$ so that $dt = dx$. Adding a zero valued term $\int_{t=0}^{a} e^{-st}.0.\,dt$ on the RHS of Eq. (2.6), we get

$$e^{-as}.f(s) = \int_{t=0}^{a} e^{-st}.0.\,dt + \int_{x=a}^{\infty} e^{-sx}.F(x-a).\,dx$$

$$= \int_{t=0}^{a} e^{-st}.0.\,dt + \int_{t=a}^{\infty} e^{-st}.F(t-a)\}\,dt$$

$$= \int_{t=0}^{\infty} e^{-st}.G(t).\,dt = £G(t) \Rightarrow \text{Eq. (2.5). //}$$

Note 2.1. The result given by Eq. (2.5) can also be written as

$$£^{-1}\{e^{-as}.f(s)\} = F(t-a).\,U(t-a), \qquad (2.7)$$

where $U(t-a)$ is Heaviside's unit function defined by Eq. (6.10.34).

Theorem 2.4 (*Change of scale property*). If $£^{-1}\{f(s)\} = F(t)$, then

$$£^{-1}\{f(a.s)\} = \{F(t/a)\}/a. \qquad (2.8)$$

Proof. Replacing s by as in Eq. (6.1.5), and putting $a\,t = x \Rightarrow dt = (dx)/a$, we get

$$f(a.s) = \int_{t=0}^{\infty} e^{-a.s\,t}.\,F(t).\,dt = (1/a)\int_{x=0}^{\infty} e^{-s.x}.\,F(x/a).\,dx$$

$$= (1/a)\int_{t=0}^{\infty} e^{-s.t}.\,F(t/a).\,dt = (1/a).\,£\,F(t/a).$$

Operating it by $£^{-1}$, we derive Eq. (2.8). //

Example 2.1. Find the inverse Laplace transform of

$$f(s) = (s-2)/\{(s-2)^2+5^2\} + (s+4)/\{(s+4)^2+9^2\}$$
$$+ 1/\{(s+2)^2+3^2\}.$$

Solution. Applying Eqs. (2.2) and (2.4), the inverse Laplace transform of $f(s)$ is

$$e^{2t}.\,£^{-1}\{s/(s^2+5^2)\} + e^{-4t}.\,£^{-1}\{s/(s^2+9^2)\} + e^{-2t}.\,£^{-1}\{1/(s^2+3^2)\}$$

$$= e^{2t}.\cos 5t + e^{-4t}.\cos 9t + (e^{-2t}.\sin 3t)/3. \ //$$

Example 2.2. Evaluate $£^{-1}\{(s+1)/(s^2+6s+25)\}$.

Solution. Rewriting the function as

$$f(s) = (s+3)/\{(s+3)^2+4^2\} - 2/\{(s+3)^2+4^2\},$$

and applying Eq. (2.4), the inverse Laplace transform of $f(s)$ is

$$£^{-1}[(s+3)/\{(s+3)^2+4^2\}] - 2.\,£^{-1}[1/\{(s+3)^2+4^2\}]$$

$$= e^{-3t}.\,£^{-1}\{s/(s^2+4^2)\} - 2.\,e^{-3t}.\,£^{-1}\{4/(s^2+4^2)\}/4$$

$$= e^{-3t}.\,\{\cos 4t - (\sin 4t)/2\}. \ //$$

Example 2.3. Evaluate $£^{-1}\{e^{-3s}/(s-4)^2\}$.

Solution. Let $$f(s) = 1/(s-4)^2 = £\,F(t), \tag{2.9}$$

so that

$$F(t) = £^{-1}\{f(s)\} = £^{-1}\{1/(s-4)^2\} = e^{4t}.\,£^{-1}(1/s^2) = e^{4t}.t, \tag{2.10}$$

by Eqs. (2.2) and (6.2.2). Therefore, by Theorem 2.3 and Eq. (2.10),

$$£^{-1}\{e^{-3s}f(s)\} = £^{-1}\{e^{-3s}/(s-4)^2\} = F(t-3)$$

$$= (t-3).\,e^{4(t-3)} \text{ when } t \geq 3, \quad \text{or} \quad 0, \text{ when } t < 3$$

$$= (t-3).\,e^{4(t-3)}.\,U(t-3),$$

where $U(t-3)$ is the Heaviside's unit function.

Alternately, the function in Eq. (2.9), in view of the results given by Eqs. (6.2.4) and (6.7.1a), may also be written as

$$f(s) = 1/(s-4)^2 = -(d/ds)\{1/(s-4)\} = -(d/ds)(£\,e^{4t})$$

$$= £\,(t.\,e^{4t}) = £\,F(t).$$

This yields Eq. (2.10). Rest follows as above. //

§ 3. Methods to Find Inverse Laplace Transforms

3.1. Partial fractions method: The functions are split into partial fractions and the operator $£^{-1}$ is then carried out on the individual fractioned terms. The method is demonstrated by means of the following example.

Example 3.1. Evaluate $£^{-1}\{6/(2s-3) + (7s+6)/(s^2+s-6)\}$.

Solution. Resolving the second fraction into partial fractions, we have

$$(7s+6)/(s^2+s-6) = (7s+6)/(s-2)(s+3) = 4/(s-2) + 3/(s+3).$$

Therefore, the desired inverse Laplace transform is

$$3.\ £^{-1}\{1/(s-3/2)+4.\ £^{-1}\{1/(s-2)\}+3.\ £^{-1}\{1/(s+3)\}$$

$$= 3e^{3t/2}+4e^{2t}+3e^{-3t}, \qquad \text{by Eq. (6.2.4).} \ //$$

3.2. Method of differentiation w.r.t. some parameter: The method is demonstrated by means of the following examples.

Example 3.2. Evaluate $£^{-1}\{s/(s^2+a^2)^2\}$.

Solution. The given function $f(s) = s/(s^2+a^2)^2$ contains a parameter a. We compute the derivative of the function $s/(s^2+a^2)$ w.r.t. the parameter a:

$$(d/da)\{s/(s^2+a^2)\} = -2a\,s/(s^2+a^2)^2 = -2a.\{s/(s^2+a^2)^2\} \qquad (3.1)$$

$$\Rightarrow \quad £^{-1}[(d/da)\{s/(s^2+a^2)\}] = -2a.\ £^{-1}\{s/(s^2+a^2)^2\}$$

$$\Rightarrow \quad £^{-1}\{s/(s^2+a^2)^2\} = -(1/2a).\ £^{-1}[(d/da)\{s/(s^2+a^2)\}]$$

$$= -\ (1/2a)\,(d/da)\,[£^{-1}\{s/(s^2+a^2)\}]$$

$$= -\ (1/2a)\,(d/da)\,(\cos at) = (t/2a).\ \sin at, \qquad \text{by Eq. (6.2.5a).}$$

Method 2. In view of the result (6.2.5a), the LHS of (3.1) may also be written as

$$(d/da)\,(£\cos at) = £\{(d/da)(\cos at)\} = £\{-t.\ \sin at\}.$$

Thus, (3.1) becomes

$$£\{-t.\ \sin at\} = -2a.s/(s^2+a^2)^2.$$

Operating it by $(-1/2a)\,£^{-1}$, we immediately get the desired result.

Method 3. For $s/(s^2+a^2)^2 = (1/2a)\{-(d/ds)\{a/(s^2+a^2)\}$

$$= (1/2a)\,[-(d/ds)\{£(\sin at)\}] = (1/2a)\,£(t.\sin at), \quad \text{by Eq. (6.7.1a)}$$

$$\Rightarrow \qquad £^{-1}\{s/(s^2+a^2)^2\} = (t/2a).\ \sin at. \ // \qquad (3.2)$$

Example 3.3. Show that

$$£^{-1}\{s/(s^4+s^2+1)\} = (2/\sqrt{3}).\ \sinh(t/2).\ \sin(t\sqrt{3}/2).$$

Solution. Factorizing s^4+s^2+1 as

$$s^4+s^2+1 = (s^2+1)^2-s^2 = (s^2+1+s).(s^2+1-s),$$

and breaking the given function into partial fractions:

$$s/(s^4+s^2+1) = (as+b)/(s^2+s+1)+(cs+d)/(s^2-s+1), \quad (3.3)$$

where a, b, c and d are some constants to be determined. Multiplying on both sides by $(s^2+s+1).(s^2-s+1)$, we have

$$s = (as+b)(s^2-s+1)+(cs+d)(s^2+s+1)$$

$$= (a+c)s^3+(-a+b+c+d)s^2+(a-b+c+d)s+(b+d).$$

Equating the coefficients of like terms, we get

$$a+c = 0, -a+b+c+d = 0, a-b+c+d = 1, b+d = 0.$$

Solving these simultaneous equations, we conclude $a = c = 0$, $b = -1/2$, $d = 1/2$; which reduce the RHS of (3.3) to

$$-1/2\,(s^2+s+1)+1/2\,(s^2-s+1)$$

$$= -1/2\ \{(s+1/2)^2+(\sqrt{3}/2)^2\} + 1/2\ \{(s-1/2)^2+(\sqrt{3}/2)^2\}.$$

Hence, the desired result reads as

$$-(1/2)\,£^{-1}[1/\{(s+1/2)^2+(\sqrt{3}/2)^2\}]+(1/2)\,£^{-1}[1/\{(s-1/2)^2+(\sqrt{3}/2)^2\}]$$

$$= -(1/2)\,e^{-t/2}\,£^{-1}[1/\{s^2+(\sqrt{3}/2)^2\}]$$

$$+(1/2)\,e^{t/2}.\,£^{-1}[1/\{s^2+(\sqrt{3}/2)^2\}], \qquad \text{by Eqs. (2.2) and (2.4),}$$

$$= \{(e^{t/2}-e^{-t/2})/2\}.\,£^{-1}[1/\{s^2+(\sqrt{3}/2)^2\}]$$

$$= (2/\sqrt{3}).\ \sinh(t/2).\ \sin(t\sqrt{3}/2), \qquad \text{by Eq. (6.2.5b).} //$$

§ 4. Inverse Laplace Transforms of Derivatives and Integrals

Theorem 4.1. If $£^{-1}\{f(s)\} = F(t)$, then

$$£^{-1}\{f^{(n)}(s)\} = (-1)^n.\, t^n.\, F(t) = (-1)^n.\, t^n.\, £^{-1}\{f(s)\}, \quad (4.1)$$

where $n = 1, 2, 3, \ldots$; and $f^{(n)}$ denotes the nth order derivative of $f(s)$ w.r.t. s.

Proof. The Laplace transform of $t^n.\, F(t)$ is given by Eq. (6.7.1), that yields

$$t^n.\, F(t) = £^{-1}\{(-1)^n f^{(n)}(s)\} = (-1)^n.\, £^{-1}\{f^{(n)}(s)\}.$$

Dividing it by the coefficient $(-1)^n$, we immediately get Eq. (4.1).

Particularly, for $n = 1$, the result given by Eq. (4.1) reads as

$$£^{-1}\,(d/ds)\,\{f(s)\} = -t.\, F(t) = -\,t.\, £^{-1}\{f(s)\}. \,// \quad (4.2)$$

Theorem 4.2. If $£^{-1}\{f(s)\} = F(t)$, then

$$£^{-1}\,[\{f(s)\}/s] = \int_{u=0}^{t} F(u)\,du. \quad (4.3)$$

Proof. The Laplace transform of an integral function is derived vide equation (6.6.2), that immediately yields Eq. (4.3). //

Example 4.1. Find $£^{-1}\{1/s^2(s+1)^2\}$.

Solution. Replacing a by $-a$ in Eq. (6.2.4), we have

$$£\,(e^{-at}) = 1/(s+a) \Rightarrow £^{-1}\{1/(s+1)\} = e^{-t}. \quad (4.4)$$

Applying Eq. (4.1), for $n = 1$, to the function $f(s) \equiv 1/(s+1)$, we derive

$$£^{-1}[(d/ds)\,\{1/(s+1)\}] = -t.\, £^{-1}\{1/(s+1)\} = -t\,e^{-t}, \text{ by Eq. (4.4);}$$

or, $$£^{-1}\{1/(s+1)^2\} = t.\, e^{-t} \equiv F(t), \text{ say.} \quad (4.5)$$

Next, applying Theorem 4.2 to the function $1/(s+1)^2$, we obtain

$$£^{-1}[\{1/s\,(s+1)^2\} = \int_{u=0}^{t} u.\,e^{-u}.\,du = \left[-u.e^{-u}\right]_0^t + \int_{u=0}^{t} e^{-u}.\,du$$

$$= -t.\,e^{-t} - \left[e^{-u}\right]_0^t = -t.\,e^{-t} - e^{-t} + 1 = 1-(t+1).\,e^{-t}. \quad (4.6)$$

Applying Theorem 4.2 again to the function $1/s\,(s+1)^2$, we derive

$$£^{-1}[\{1/s^2(s+1)^2\} = \int_{u=0}^{t} \{1-(u+1).\,e^{-u}\}.\,du$$

$$= \left[u+(u+1).e^{-u}\right]_0^t - \int_{u=0}^{t} e^{-u}.\,du$$

$$= t+(t+1).\,e^{-t} - 1 + \left[e^{-u}\right]_0^t = t+(t+1).\,e^{-t} - 1 + e^{-t} - 1$$

$$= (t+2).\,e^{-t} + t - 2.$$

Alternately, breaking the given function into partial fractions:

$$1/s^2(s+1)^2 = a/s + b/s^2 + c/(s+1) + d/(s+1)^2 \quad (4.7)$$

$$\Rightarrow \quad 1 = a.\,s\,(s+1)^2 + b.\,(s+1)^2 + c.\,s^2\,(s+1) + d.\,s^2.$$

Giving 0, – 1, 1 and – 2 values to s in above identity and equating the coefficients of like terms we derive

$$b = d = 1, \quad 4a+2c = -4 \Rightarrow 2a+c = -2 \quad \text{and} \quad 2a+4c = 4.$$

The last two relations determine the remaining coefficients $a = -2$ and $c = 2$. Thus, putting for the coefficients a, b, c, d and operating by $£^{-1}$ in Eq. (4.7), we find

$$£^{-1}\{1/s^2(s+1)^2\} = -2\,£^{-1}(1/s) + £^{-1}(1/s^2)$$

$$+ 2\,£^{-1}\{1/(s+1)\} + £^{-1}\{1/(s+1)^2\}.$$

The first three terms on the RHS can be evaluated by means of Eqs. (6.2.1), (6.2.2) and (6.2.4); whereas the last term alters as

$$£^{-1}[-(d/ds)\{1/(s+1)\}] = £^{-1}[-(d/ds)\{£\,e^{-t}\}], \quad \text{by Eq. (6.2.4)}$$

$$= \pounds^{-1}\{\pounds(t\,e^{-t})\} = t\,e^{-t}, \qquad \text{by (6.7.1a).}$$

Therefore, the desired result is

$$\pounds^{-1}\{1/s^2(s+1)^2\} = -2+t+2\,e^{-t}+t\,e^{-t} = (t+2)\,e^{-t}+t-2.\ //$$

Example 4.2. Find the inverse Laplace transform of the function

$$f(s) \equiv (2s+1)/(s+2)^2(s-1)^2.$$

Solution. Breaking the function into partial fractions:

$$f(s) = a/(s+2)+b/(s+2)^2+c/(s-1)+d/(s-1)^2.$$

Multiplying on both sides by $(s+2)^2(s-1)^2$, we have

$$2s+1 = a(s+2)(s-1)^2+b(s-1)^2+c(s-1)(s+2)^2+d.(s+2)^2. \tag{4.8}$$

Choosing $s = -2, 1, 0, -1$, we evaluate the coefficients a, b, c, d:

$$b = -1/3, \quad d = 1/3, \quad 2a-4c = 1+1/3-4/3 = 0,$$

and

$$4a-2c = -1+4/3-1/3 = 0 \Rightarrow a = c = 0.$$

Hence, the desired partial fractions are

$$f(s) = (1/3)\{1/(s-1)^2-1/(s+2)^2\}$$

$\Rightarrow$

$$\pounds^{-1}\{f(s)\} = (1/3)[\pounds^{-1}\{1/(s-1)^2\}-\pounds^{-1}\{1/(s+2)^2\}] \tag{4.9}$$

$$= (1/3)[-\pounds^{-1}(d/ds)\{1/(s-1)\}+\pounds^{-1}(d/ds)\{1/(s+2)\}]$$

$$= (1/3)[t.\pounds^{-1}\{1/(s-1)\}-t.\pounds^{-1}\{1/(s+2)\}], \qquad \text{by Eq. (4.2)}$$

$$= t.(e^{t}-e^{-2t})/3, \qquad \text{by Eq. (6.2.4).}$$

Alternately, applying the first shifting property given by Eq. (2.2), the RHS of Eq. (4.9) can also be evaluated as

$$(1/3)\{e^{t}.\pounds^{-1}(1/s^2)-e^{-2t}\pounds^{-1}(1/s^2)\} = t(e^{t}-e^{-2t})/3, \text{ by Eq.(6.2.4).//}$$

§ 5. Convolution Theorem

Given two piece-wise continuous functions defined over every finite interval of $t \geq 0$ their *convolution* (or generalized product) is defined by

$$F(t) * G(t) \equiv \int_{u=0}^{t} F(u) \,.\, G(t-u)\, du. \qquad (5.1)$$

Note 5.1. It may be easily verified that the convolution operator is associative, commutative and distributive over addition:

$$F*(G*H) = (F*G)*H, \qquad F*G = G*F, \qquad F*(G+H) = F*G + F*H.$$

Theorem 5.1. If $£^{-1}\{f(s)\} = F(t)$ and $£^{-1}\{g(s)\} = G(t)$, then

$$£^{-1}\{f(s).\, g(s)\} = F(t) * G(t) = \int_{u=0}^{t} F(u).\, G(t-u)\, du. \quad (5.2)$$

Proof. By definition of Laplace transform,

$$£\{F(t) * G(t)\} = \int_{t=0}^{\infty} e^{-st}\{F(t) * G(t)\}\, dt$$

$$= \int_{t=0}^{\infty} e^{-st}\, \{\int_{u=0}^{t} F(u).G(t-u)\, du\}.\, dt, \qquad (5.3)$$

by Eq. (5.1). In the integral within curly brackets u varies from 0 to t covering the entire region of integration that lies between $t = 0$ and $t = \infty$. When the order of integration in above double integration is changed the limits of integration in the changed integral (within curly brackets) become $t = u$ to $t = \infty$; and the area of integration lies between $u = 0$ to $u = \infty$. Hence, changing the order of integration, the equation (5.3) reduces to

$$£\{F(t) * G(t)\} = \int_{u=0}^{\infty} F(u).\, \{\int_{t=u}^{\infty} e^{-st}.\, G(t-u)\, dt\}\, du.$$

Putting $t - u = v$ so that $dt = dv$, above integral reduces to

$$\int_{u=0}^{\infty} F(u).\, \{\int_{v=0}^{\infty} e^{-s(u+v)}.\, G(v)\, dv\}.\, du$$

$$= \int_{u=0}^{\infty} F(u)\, e^{-su}.\, \{\int_{v=0}^{\infty} e^{-sv}.\, G(v)\, dv\}.\, du$$

$$= \int_{u=0}^{\infty} F(u).\, e^{-su}.\{£\, G(t)\}.\, du = \int_{u=0}^{\infty} F(u).\, e^{-su}.\, g(s)\}\, du$$

$$= \{\int_{u=0}^{\infty} F(u).\, e^{-su}.\, du\}.\, g(s) = \{\, £\, F(t)\,\}.\, g(s)$$

$$= f(s).\, g(s) = \{\, £\, F(t)\,\}\, \{\, £\, G(t)\,\}. \qquad (5.4)$$

Operating it $£^{-1}$, there results the equation (5.2).

Combining Eqs. (5.2) and (5.4), we may also have

$$£\,\{\, F(t) * G(t)\} = £\,\{\int_{u=0}^{t} F(u).\, G(t-u)\, du\,\}$$

$$= f(s).\, g(s) = \{\, £\, F(t)\,\}.\{\, £\, G(t)\,\}. // \qquad (5.5)$$

Example 5.1. Using the *convolution theorem* evaluate the following:

(i) $£^{-1}\{1/s^2(s+1)^2\}$, **(ii)** $£^{-1}\{\, s/(s^2+a^2)^2\}$.

Solution. (i) From Eq. (6.2.2), we have

$$£^{-1}\,\{1/s^2) = t \equiv G(t) \quad \Rightarrow \quad t-u \equiv G(t-u). \qquad (5.6)$$

Also, Eq. (4.5) determines the inverse Laplace transform of $1/(s+1)^2$. Therefore, Eq. (5.2) gives

$$£^{-1}\,[\,\{\, 1/(s+1)^2\}\,(1/s^2)\,] = \int_{u=0}^{t} (u.\, e^{-u})\,(t-u)\, du$$

$$= \int_{u=0}^{t} (u\,.t - u^2).\, e^{-u}.\, du.$$

Integrating it repeatedly by *method of parts* taking e^{-u} as the second function, above integral simplifies to

$$\left[(-ut+u^2)\,e^{-u}\right]_0^t + \int_{u=0}^{t} (t-2u)\, e^{-u}.\, du$$

$$= \left[(-t+2u)\,e^{-u}\right]_0^t - 2\int_{u=0}^{t} e^{-u}.\, du$$

$$= t\,e^{-t} + t + 2\left[e^{-u}\right]_0^t = t\,e^{-t} + t + 2\,(e^{-t}-1) = (t+2)\,e^{-t} + t - 2$$

as in Example 4.1.

(ii) The results given by equations (6.2.5) yield

$$£^{-1}\{s/(s^2+a^2)\} = \cos at \equiv F(t),$$

$$£^{-1}\{1/(s^2+a^2)\} = (\sin at)/a \equiv G(t), \qquad \text{say.}$$

Hence, an application of *convolution theorem* for these functions determines

$$£^{-1}[\{s/(s^2+a^2)\}.\{1/(s^2+a^2)\}]$$

$$= (1/a)\int_{u=0}^{t} \cos au.\ \sin a(t-u).\ du \qquad (5.7)$$

$$= (1/a)\int_{u=0}^{t} \cos au.\ (\sin at.\ \cos au - \cos at.\ \sin au).\ du$$

$$= (1/a).\{\sin at.\int_{u=0}^{t} \cos^2 au.\ du - \cos at.\int_{u=0}^{t} \cos au.\ \sin au.\ du\}$$

$$= (1/2a).\{\sin at.\int_{u=0}^{t} (1+\cos 2au).\ du - \cos at.\int_{u=0}^{t} \sin 2au.\ du\}$$

$$= (1/2a).\{\sin at.\left[u+(\sin 2au)/2a\right]_0^t + \cos at.\ \left[(\cos 2au)/2a\right]_0^t\}$$

$$= (1/2a).[\sin at.\{t+(\sin 2at)/2a\} + \cos at.\ \{(\cos 2at-1)/2a\}]$$

$$= (t.\ \sin at)/2a + (\sin at.\ \sin 2at + \cos at.\ \cos 2at - \cos at)/4a^2$$

$$= (t.\ \sin at)/2a + \{\cos(2at-at) - \cos at\}/4a^2 = (t.\ \sin at)/2a, \quad (5.8)$$

as in Example 3.2. //

Example 5.2. Using the *convolution theorem* establish the identity

$$\int_{u=0}^{t} \sin u.\ \cos(t-u).\ du = (t.\ \sin t)/2. \qquad (5.9)$$

Solution. Taking $a = 1$ in the Example 5.1(ii), and applying the *convolution theorem* to the functions

$$£^{-1}\{s/(s^2+1)\} = \cos t \equiv G(t) \qquad (5.10)$$

$\Rightarrow$

$$\cos(t-u) = G(t-u),$$

and

$$£^{-1}\{1/(s^2+1)\} = \sin t \equiv F(t), \qquad (5.11)$$

we get

$$£^{-1}[\{1/(s^2+1)\}\{s/(s^2+1)\}] = \int_{u=0}^{t} \sin u.\cos(t-u).du.$$

Proceeding similarly as in previous Example, the integral simplifies to

$$\int_{u=0}^{t} \sin u.(\cos t.\cos u + \sin t.\sin u).\,du$$

$$= \{\cos t.\int_{u=0}^{t} \sin u.\cos u.\,du + \sin t.\int_{u=0}^{t} \sin^2 u.\,du\}$$

$$= \{\cos t.\int_{u=0}^{t} \sin 2u.\,du + \sin t.\int_{u=0}^{t} (1-\cos 2u)\,du\}/2$$

$$= \{\cos t.\left[-(\cos 2u)/2\right]_0^t + \sin t.\left[u-(\sin 2u)/2\right]_0^t\}/2$$

$$= [\cos t.(1-\cos 2t)/2 + \sin t.\{t-(\sin 2t)/2\}]/2$$

$$= \{\cos t - (\cos t.\cos 2t + \sin t.\sin 2t)\}/4 + (t.\sin t)/2$$

$$= \{\cos t - \cos(2t-t)\}/4 + (t.\sin t)/2 = (t.\sin t)/2.$$

Alternately, in view of (5.5), we have

$$£\,H(t) \equiv £\,\{\int_{u=0}^{t} \sin u.\cos(t-u).\,du\} = (£\sin t).(£\cos t)$$

$$= \{1/(s^2+1)\}\{s/(s^2+1)\} = s/(s^2+1)^2 = (-1/2)(d/ds)\{1/(s^2+1)\}$$

$$\Rightarrow \qquad H(t) = (-1/2)\,£^{-1}[(d/ds)\{1/(s^2+1)\}]$$

$$= (t/2).\,£^{-1}\{1/(s^2+1)\} = (t.\sin t)/2,$$

by Eqs. (4.2) and (5.11). //

§ 6. Problem Set

6.1. Find the inverse Laplace transforms of the following functions:

(i) $(s^2-3s+4)/s^3$, **(ii)** $(s+23)/(s^2-4s+13)$,

(iii) $(2s^2-6s+5)/(s^3-6s^2+11s-6)$, **(iv)** $(4s+5)/(s-1)^2(s+2)$,

(v) $(5s + 3) / (s - 1)(s^2 + 2s + 5)$, **(vi)** $s / (s^4 + 4a^4)$, **(vii)** $s^3 / (s^4 - a^4)$,

(viii) $1/ (s^3 - a^3)$, **(ix)** $(s^2 + 6) / (s^2 + 1)(s^2 + 4)$, **(x)** $s / (s + 1)^2.(s^2 + 1)$.

[Hint: (i) The function splits as

$$f(s) \equiv 1/s - 3/s^2 + 4/s^3 \Rightarrow £^{-1} f(s) = 1 - 3t + 2t^2,$$

by Eqs. (6.2.1) – (6.2.3).

(ii) The function splits as

$$f(s) \equiv (s - 2) / \{(s - 2)^2 + 3^2\} + 25 / \{(s - 2)^2 + 3^2\}$$

$$\Rightarrow$$

$$£^{-1} f(s) = e^{2t} [£^{-1}\{s / (s^2 + 3^2)\} + (25/3)\, £^{-1}\{3/\{(s^2 + 3^2)\}\}], \text{ by Eq. (2.2)}$$

$$= e^{2t} \{\cos 3t + (25/3) \sin 3t\}, \qquad \text{by Eq. (6.2.5).}$$

(iii) $$f(s) \equiv 1 / 2(s - 1) - 1 / (s - 2) + 5 / 2(s - 3)$$

$$\Rightarrow$$

$$£^{-1} f(s) = (1/2) [£^{-1}\{1/ (s - 1)\} - 2.£^{-1}\{1/(s - 2)\} + 5.£^{-1}\{1/(s - 3)\}$$

$$= (1/2) (e^{t} - 2e^{2t} + 5e^{3t}), \qquad \text{by Eq. (6.2.4).}$$

(iv) $$f(s) \equiv (1/3)\{1 / (s - 1) - 1 / (s + 2)\} + 3 / (s - 1)^2$$

$$\Rightarrow$$

$$£^{-1} f(s) = (1/3) [£^{-1} \{1 / (s - 1)\} - £^{-1} \{1 / (s + 2)\}$$

$$- 3\, £^{-1} [(d/ds) \{1 / (s - 1)\}]$$

$$= (1/3) (e^{t} - e^{-2t}) + 3.t.\, £^{-1} \{1 / (s - 1)\}, \quad \text{by Eqs. (6.2.4) and (4.2)}$$

$$= (1/3) (e^{t} - e^{-2t}) + 3.t.\, e^{t}.$$

(v) $$f(s) \equiv 1 / (s - 1) + (- s + 2) / \{(s + 1)^2 + 2^2\}$$

$$= 1 / (s - 1) - (s + 1) / \{(s + 1)^2 + 2^2\} + 3 / \{(s + 1)^2 + 2^2\}$$

$$\Rightarrow$$

$$£^{-1} f(s) = e^{t} - e^{-t} [£^{-1} \{s / (s^2 + 2^2)\} - (3/2)\, £^{-1}\{2 / (s^2 + 2^2)\}],$$

by Eqs. (6.2.4) and (2.2); and so is

$= e^{t} - e^{-t}\{\cos 2t - (3/2)\sin 2t\},$ by (6.2.5).

(vi) $f(s) \equiv s/(s^2 - 2as + 2a^2).(s^2 + 2as + 2a^2)$

$= (1/4a)\{1/(s^2 - 2as + 2a^2) - 1/(s^2 + 2as + 2a^2)\}$

$\Rightarrow$

$£^{-1}f(s) = (1/4a)\, £^{-1}[1/\{(s-a)^2 + a^2\} - 1/\{(s+a)^2 + a^2\}]$

$= (1/4a)\,[e^{at}.£^{-1}\{1/(s^2 + a^2)\} - e^{-at}.£^{-1}\{1/(s^2 + a^2)\}],$ by Eq. (2.2)

$= (1/4a^2)(e^{at} - e^{-at}).£^{-1}\{a/(s^2 + a^2)\} = (1/4a^2)(e^{at} - e^{-at}).\sin at,$

by Eq. (6.2.5)

$= (1/2a^2)\sinh at.\ \sin at.$

(vii) $f(s) \equiv s^3/(s^2 - a^2)(s^2 + a^2) = (1/2)\{s/(s^2 - a^2) + s/(s^2 + a^2)\}$

$\Rightarrow$

$£^{-1}f(s) = £^{-1}\{s/(s^2 - a^2) + s/(s^2 + a^2)\}/2 = (\cosh at + \cos at)/2,$

by (6.2.5a) and (6.2.6a).

(viii) $f(s) \equiv 1/(s-a)(s^2 + as + a^2)$

$= (1/3a^2)\{1/(s-a) - (s+2a)/(s^2 + as + a^2)\}$

$\Rightarrow$

$£^{-1}f(s) = (1/3a^2)\, £^{-1}\{1/(s-a)\}$

$- (1/3a^2)\, £^{-1}[(s + a/2 + 3a/2)/\{(s + a/2)^2 + 3a^2/4\}]$

$= (1/3a^2)\, e^{at} - (1/3a^2)\, e^{-at/2}.\,[£^{-1}\{s/(s^2 + 3a^2/4)\}$

$+ £^{-1}\{(3a/2)/\{(s^2 + 3a^2/4)\}],$ by (2.2)

$= (1/3a^2)\,[e^{at} - e^{-at/2}.\{\cos(at\sqrt{3}/2) + \sqrt{3}.\sin(at\sqrt{3}/2)\}],$ by (6.2.5).

(ix) The functions splits as $\{5/(s^2+1) - 2/(s^2+4)\}/3$. Hence, its inverse Laplace transform, by (6.2.5b), can be written as $(5\sin t - \sin 2t)/3$. //]

6.2. Find the inverse Laplace transforms of the following functions by using their different properties:

(i) $1/(s^2+a^2)$, **(ii)** $1/s(s+1)^3$, **(iii)** $\ln\{(s+1)/(s-1)\}$,

(iv) $\ln\{s.(s+1)/(s^2+4)\}$, **(v)** $\cot^{-1}(s/a)$, **(vi)** $s^2/(s^2+a^2)^2$,

(vii) $1/(s^2+a^2)^2$, **(viii)** $\tan^{-1}(2/s)$, **(ix)** $\ln\{(s^2+a^2)/(s^2+b^2)\}$,

(x) $\ln(1-a^2/s^2)$, **(xi)** $\ln\{(s+a)/(s+b)\}$.

[Hint: (ii) $f(s) \equiv 1/s - 1/(s+1) - 1/(s+1)^2 - 1/(s+1)^3$

$\Rightarrow$

$$£^{-1}f(s) = £^{-1}(1/s) - £^{-1}\{1/(s+1)\} - £^{-1}\{1/(s+1)^2\} - £^{-1}\{1/(s+1)^3\}$$

$$= 1 - e^{-t} + £^{-1}\{(d/ds)(s+1)^{-1}\} + (1/2)\,£^{-1}\{(d/ds)(s+1)^{-2}\}$$

$$= 1 - e^{-t} - t.\,£^{-1}\{(s+1)^{-1}\} - (t/2).\,£^{-1}\{(s+1)^{-2}\}, \text{ by (4.2)}$$

$$= 1 - e^{-t} - t.\,e^{-t} + (t/2)\,£^{-1}\{(d/ds)(s+1)^{-1}\}$$

$$= 1 - e^{-t} - t\,e^{-t} - (t/2)\,t.\,£^{-1}\{(s+1)^{-1}\} = 1 - e^{-t} - t.\,e^{-t} - (t^2 e^{-t})/2.$$

(iii) $f(s) \equiv \ln\{(s+1)/(s-1)\} = \ln(s+1) - \ln(s-1)$

so that

$$df(s)/ds = 1/(s+1) - 1/(s-1).$$

Its inverse Laplace transform, for (6.2.4) and (4.2), yields

$$-t.\,F(t) = e^{-t} - e^{t} \quad \Rightarrow \quad F(t) \equiv £^{-1}\{f(s)\} = (e^{t} - e^{-t})/t.$$

(iv) $f(s) \equiv \ln\{s.(s+1)/(s^2+4)\} = \ln s + \ln(s+1) - \ln(s^2+4)$

so that

$$df(s)/ds = 1/s + 1/(s+1) - 2s/(s^2+4).$$

Its inverse Laplace transform, for (6.2.1), (6.2.4), (6.2.5a) and (4.2), yields

$$-t.\,F(t) = 1 + e^{-t} - 2\cos 2t$$

$\Rightarrow$

$$F(t) \equiv £^{-1}\{f(s)\} = -(1 + e^{-t} - 2\cos 2t)/t.$$

(v) $f(s) \equiv \cot^{-1}(s/a) \Rightarrow df(s)/ds = -a/(s^2+a^2)$.

Its inverse Laplace transform, for (6.2.5b) and (4.2), yields

$$t.\,F(t) = \sin at \quad \Rightarrow \quad F(t) \equiv £^{-1}\{f(s)\} = (\sin at)/t.$$

(vi) The inverse Laplace transform of a function $f(s) \equiv s/(s^2+a^2)^2$ is evaluated in Example 3.2:

$$£^{-1}\{f(s)\} = £^{-1}\{s/(s^2+a^2)^2\} = (t/2a).\,\sin at \equiv F(t), \text{ say}$$

$\Rightarrow$

$$F(0) = 0, \text{ and } F'(t) = (1/2a).\,\sin at + (t/2).\,\cos at. \qquad (6.1)$$

Hence (6.5.1), when operated by $£^{-1}$, determines

$$£^{-1}\{s.f(s)\} \equiv £^{-1}\{s^2/(s^2+a^2)^2\}$$

$$= F'(t) = (1/2a).\,\sin at + (t/2).\,\cos at, \qquad \text{by (6.1).}$$

Alternately, $f(s) \equiv s^2/(s^2+a^2)^2 = 1/(s^2+a^2) - a^2/(s^2+a^2)^2$

$$= 1/(s^2+a^2) + (a/2)\,(d/da)\,\{1/(s^2+a^2)\}$$

$\Rightarrow$

$$£^{-1} f(s) = £^{-1}\{1/(s^2+a^2)\} + (a/2)\,(d/da)\,£^{-1}\{1/(s^2+a^2)\}$$

$$= (\sin at)/a + (a/2)\,(d/da)\,\{(\sin at)/a\}, \qquad \text{by (6.2.5b)}$$

$$= (\sin at)/a + (a/2)\,\{(t/a).\,\cos at - (\sin at)/a^2\}$$

$$= (\sin at)/2a + (t/2).\,\cos at.$$

(vii) $f(s) \equiv 1/(s^2+a^2)^2$

$\Rightarrow$

$$£^{-1}\{s\,f(s)\} = £^{-1}\{s/(s^2+a^2)^2\} = (t/2a).\,\sin at, \quad \text{by Eq. (3.2).}$$

Also, (6.5.1) yields,

$$£^{-1}\{s\,f(s)\} = F'(t) + £^{-1}\{F(0)\}.$$

Hence, an integral of

$$F'(t) + £^{-1}\{F(0)\} = (t/2a).\,\sin at$$

w.r.t. t determines

$$F(t) = \int (t/2a).\,\sin at.\,dt - t.\,£^{-1}\{F(0)\}$$

$$= -(t/2a^2)\cos at + (1/2a^3)\sin at - t.\,£^{-1}\{F(0)\}.$$

As $F(0) = 0$, above function simplifies to

$$F(t) \equiv £^{-1}\{1/(s^2+a^2)^2\} = (-at.\cos at + \sin at)/2a^3. \quad (6.2)$$

(viii) $f(s) \equiv \tan^{-1}(2/s) \Rightarrow df(s)/ds = -2/(s^2+4).$

Its inverse Laplace transform, for (6.2.5b) and (4.2), yields

$$t.F(t) = \sin 2t \quad \Rightarrow \quad F(t) \equiv £^{-1}\{f(s)\} = (\sin 2t)/t.$$

(ix) $f(s) \equiv \ln\{(s^2+a^2)/(s^2+b^2)\} = \ln(s^2+a^2) - \ln(s^2+b^2),$

so that

$$df(s)/ds = 2\{s/(s^2+a^2) - s/(s^2+b^2)\}.$$

Its inverse Laplace transform, for (6.2.5a) and (4.2), yields

$$-t.F(t) = 2(\cos at - \cos bt)$$

$\Rightarrow$

$$F(t) \equiv £^{-1}\{f(s)\} = 2.(-\cos at + \cos bt)/t.$$

(x) $f(s) \equiv \ln(1 - a^2/s^2) = \ln(s^2 - a^2) - 2.\ln s,$

so that

$$df(s)/ds = 2\{s/(s^2-a^2) - 1/s\}.$$

Its inverse Laplace transform, for (6.2.1), (6.2.6a) and (4.2), yields

$$-t.F(t) = 2.(\cosh at - 1) \Rightarrow F(t) \equiv £^{-1}\{f(s)\} = 2.(1 - \cosh at)/t.$$

(xi) $f(s) \equiv \ln\{(s+a)/(s+b)\} = \ln(s+a) - \ln(s+b),$

so that

$$df(s)/ds = 1/(s+a) - 1/(s+b).$$

Its inverse Laplace transform, for (6.2.4) and (4.2), yields

$$-t.F(t) = e^{-at} - e^{-bt} \Rightarrow F(t) \equiv £^{-1}\{f(s)\} = (-e^{-at} + e^{-bt})/t]. //$$

6.3. Using *Convolution theorem*, evaluate the inverse Laplace transforms of the following functions:

(i) $1/s(s^2+4)$, **(ii)** $1/(s^2+a^2)^2$, **(iii)** $1/s^2(s^2+a^2)$,

(iv) $s^2/(s^2+a^2)(s^2+b^2)$, **(v)** $1/(s+1)(s+9)^2$, **(vi)** $s/(s^2+1)(s^2+4)$.

[Hint: (i) $£^{-1}\{1/s(s^2+4)\}$

$$= \{£^{-1}(1/s)\} * [£^{-1}\{1/(s^2+4)\}], \qquad \text{by Eq. (5.2)}$$

$$= 1 * (\sin 2t)/2 \equiv (1/2)\int_{u=0}^{t} \sin 2u.\, du$$

$$= -(1/4)\left[\cos 2u\right]_{u=0}^{t} = (1-\cos 2t)/4.$$

(ii) $£^{-1}\{1/(s^2+a^2)^2\}$

$$= [£^{-1}\{a/(s^2+a^2)\}] * [£^{-1}\{a/(s^2+a^2)\}]/a^2 = (\sin at) * (\sin at)/a^2$$

$$= (1/a^2)\int_{u=0}^{t} \sin au.\, \sin a(t-u)\, du, \qquad \text{by Eq. (5.2)}$$

$$= (1/2a^2)\int_{u=0}^{t} \{\cos a(2u-t) - \cos at\}\, du$$

$$= (1/2a^2)\left[\{\sin a(2u-t)\}/2a - u.\cos at\right]_{u=0}^{t}$$

$$= \{(\sin at)/a - t.\cos at\}/2a^2.$$

(iii) $£^{-1}\{1/s^2(s^2+a^2)\} = (1/a)\{£^{-1}(1/s^2)\} * [£^{-1}\{a/(s^2+a^2)\}]$

$$= (1/a)\, t * \sin at = (1/a)\int_{u=0}^{t} (\sin au).(t-u)\, du, \quad \text{by Eq. (5.2)}$$

$$= (1/a^2)\left\{\left[(-t+u)(\cos au)\right]_{u=0}^{t} - \int_{u=0}^{t} \cos au.\, du\right\}$$

$$= (1/a^2)\left\{t - (1/a)\left[\sin au\right]_{u=0}^{t}\right\} = (at - \sin at)/a^3.$$

(iv) $£^{-1}\{s^2/(s^2+a^2)(s^2+b^2)\} = [£^{-1}\{s/(s^2+a^2)\}] * [£^{-1}\{s/(s^2+b^2)\}]$

$$= \cos at * \cos bt = (1/2)\int_{u=0}^{t} 2.\cos au.\, \cos b(t-u)\, du$$

$$= (1/2)\int_{u=0}^{t} [\cos\{(a-b)u + bt\} + \cos\{(a+b)u - bt\}]\, du$$

$$= (1/2)\left[\sin\{(a-b)u+bt\}/(a-b) + \sin\{(a+b)u-bt\}/(a+b)\right]_{u=0}^{t}$$

$$= (1/2)\{(\sin at - \sin bt)/(a-b) + (\sin at + \sin bt)/(a+b)\}$$

$$= (a.\sin at - b.\sin bt) / (a^2 - b^2).$$

(v) $£^{-1}\{ 1 / (s+1)(s+9)^2 \}$

$= [£^{-1}\{1/(s+1)\}] * [£^{-1}\{1/(s+9)^2\}] = e^{-t} * (t.\, e^{-9t})$, by Eq. (4.5)

$$= \int_{u=0}^{t} (u\, e^{-9u})\, e^{-(t-u)}\, du \;=\; e^{-t}.\int_{u=0}^{t} u\, e^{-8u}\, du$$

$$= (e^{-t}/8)\, \{ \left[-u\, e^{-8u}\right]_{u=0}^{t} + \int_{u=0}^{t} e^{-8u}\, du \}$$

$$= -(t.\, e^{-9t})/8 - (e^{-t}/64)\left[e^{-8u}\right]_{u=0}^{t} = \{e^{-t} - (8t+1).\, e^{-9t}\}/64.$$

(vi) $£^{-1}\{s/(s^2+1)(s^2+4)\} = [£^{-1}\{s/(s^2+1)\}] * [£^{-1}\{1/(s^2+4)\}]$

$$= (1/2)\cos t * \sin 2t = (1/4)\int_{u=0}^{t} 2.\cos u.\sin 2(t-u)\, du$$

$$= (1/4)\int_{u=0}^{t} \{ \sin(2t-u) + \sin(2t-3u) \}\, du$$

$$= (1/4)\, \{\left[\cos(2t-u) + (1/3)\cos(2t-3u)\right]_{u=0}^{t}$$

$= (1/4)\{(\cos t - \cos 2t) + (1/3)(\cos t - \cos 2t)\} = (1/3)(\cos t - \cos 2t)]$.//

6.4. Show that:

(i) $£^{-1}\{(1/s).\sin(1/s)\} = t - t^3/(3!)^2 + t^5/(5!)^2 - t^7/(7!)^2 + \ldots \infty$,

(ii) $£^{-1}\{(1/s).\cos(1/s)\} = 1 - t^2/(2!)^2 + t^4/(4!)^2 - t^6/(6!)^2 + \ldots \infty$.

[Hint: (i) Expanding $\sin(1/s)$ in powers of $1/s$:

$$\sin(1/s) = 1/s - 1/3!\, s^3 + 1/5!\, s^5 - 1/7!\, s^7 + \ldots \infty,$$

so that

$$(1/s)\sin(1/s) = 1/s^2 - 1/3!\, s^4 + 1/5!\, s^6 - 1/7!\, s^8 + \ldots \infty.$$

Now operating by $£^{-1}$ and applying the formula (6.2.3) we get the desired result.

(ii) Proceed similarly as above.]

CHAPTER 8

APPLICATIONS OF LAPLACE TRANSFORMS TO DIFERENTIAL EQUATIONS

§ 1. Introduction

An ordinary linear differential equation with constant coefficients can be solved by using the techniques of Laplace transforms. The technique is especially useful in solving the non-homogeneous equations resulting in the modeling systems which involve discontinuous periodic input functions.

An attractive feature of using Laplace transforms for the solution of differential equations is that the transformed equation becomes an algebraic equation. To achieve the same we consider the Laplace transforms of the differential equation. The transformed equation (which is an algebraic equation in the new variable) is then solved for this variable. Finally, applying the inverse Laplace transform to this solution the original variables are retrieved.

§ 2. Solution of Ordinary Differential Equations with Constant Coefficients by Laplace Transform Methods

Let

$$d^2 y / dt^2 + a\,(dy/dt) + b\,y = F(t) \tag{2.1}$$

be an ordinary differential equation of second order with constant coefficients a and b. Considering the Laplace transform of Eq. (2.1), we get

$$£\, y''(t) + a\, £\, y'(t) + b\, £\, y(t) = £\, F(t), \tag{2.2}$$

where primes denote derivatives w.r.t. t. Applying Eqs. (6.5.1) and (6.5.2), we obtain

$$£\, y'(t) = s\, £\, y(t) - y(0), \tag{2.3}$$

$$£\, y''(t) = s^2\, £\, y(t) - s\, y(0) - y'(0). \tag{2.4}$$

Putting these values in Eq. (2.2) we, thus, derive an algebraic form of the differential equation in the variable s:

$$\{s^2\, £\, y(t) - s\, y(0) - y'(0)\} + a\, \{s\, £\, y(t) - y(0)\} + b\, £\, y(t) = £\, F(t),$$

or,

$$(s^2 + a\,s + b)\,£\,y\,(t) \;=\; (s + a)\,y\,(0) + y\,'\,(0) + £\,F\,(t)$$

$\Rightarrow$

$$£\,y\,(t) = \{(s + a)\,y\,(0) + y\,'\,(0) + F\,(t)\}\,/\,(s^2 + a\,s + b) \equiv\; y\,(s). \quad (2.5)$$

Taking its inverse Laplace transform, we obtain the solution of the given differential equation:

$$y\,(t) \;=\; £^{\,-1}\{\,y\,(s)\}. \qquad (2.6)$$

Example 2.1. Apply the Laplace transform techniques to solve the differential equation

$$d^2y\,/\,dt^2 + y \;=\; 6.\,\cos 2t, \qquad (2.7)$$

given that

$$y\,(0) = 3, \qquad y\,'\,(0) \;=\; 1. \qquad (2.8)$$

Solution. Considering the Laplace transform of Eq. (2.7), we get

$$£\,y''\,(t) + £\,y\,(t) \;=\; 6.\,£\,\cos 2t,$$

or, for Eqs. (2.4) and (6.2.5a)

$$\{s^2\,£\,y\,(t) - s\,y\,(0) - y\,'\,(0)\} + £\,y\,(t) \;=\; 6\,s\,/\,(s^2 + 4),$$

or, for Eq. (2.8)

$$(s^2 + 1)\,£\,y\,(t) \;=\; 6\,s\,/\,(s^2 + 4) \;+ 3s + 1$$

$\Rightarrow$

$$£\,y\,(t) \;=\; 6s\,/\,(s^2 + 1)\,(s^2 + 4) + 3s\,/\,(s^2 + 1) + 1/\,(s^2 + 1)$$

$$=\; 2s\,\{1/\,(s^2 + 1) - 1/\,(s^2 + 4)\} + 3s\,/\,(s^2 + 1) + 1/\,(s^2 + 1)$$

$$= 5s\,/\,(s^2 + 1) - 2s\,/\,(s^2 + 4) + 1/\,(s^2 + 1).$$

Taking its inverse Laplace transform, we obtain the solution of the given differential equation:

$$y\,(t) \;=\; 5\,£^{-1}\,\{s\,/\,(s^2 + 1)\} - 2\,£^{\,-1}\{s\,/\,(s^2 + 2^2)\} + £^{\,-1}\,\{1/\,(s^2 + 1)\}$$

$$=\; 5.\,\cos t - 2.\,\cos 2\,t + \sin t, \qquad \text{by Eq. (6.2.5).//}$$

Example 2.2. Solve the differential equation

$$d^2y\,/\,dt^2 + 9\,y \;=\; \cos 2t, \quad y\,(0) = 1, \quad y\,(\pi/2) = -\,1,$$

by using the techniques of Laplace transform.

Solution. Considering the Laplace transform of the differential equation, we get

$$£\, y''(t) + 9\, £\, y(t) = £ \cos 2t,$$

or, for Eqs. (2.4) and (6.2.5a)

$$\{s^2\, £\, y(t) - s\, y(0) - y'(0)\} + 9\, £\, y(t) = s/(s^2+4).$$

Applying the initial conditions, this equation reduces to

$$(s^2+9)\, £\, y(t) = s/(s^2+4) + s + a \qquad (2.9)$$

$\Rightarrow$

$$£\, y(t) = s/(s^2+4)(s^2+9) + s/(s^2+9) + a/(s^2+9)$$

$$= (1/5)\{s/(s^2+4) - s/(s^2+9)\} + s/(s^2+9) + a/(s^2+9)$$

$$= (1/5)\{s/(s^2+4) + 4s/(s^2+9)\} + a/(s^2+9),$$

where we have assumed $y'(0) = a$, say. Taking its inverse Laplace transform, we obtain the solution of the given differential equation:

$$y(t) = (1/5)[\, £^{-1}\{s/(s^2+2^2)\} + 4\, £^{-1}\{s/(s^2+3^2)\}]$$

$$+ (a/3)\, £^{-1}\{3/(s^2+3^2)\} = (1/5)\{\cos 2t + 4\cos 3t\} + (a/3)\sin 3t. \quad (2.10)$$

For the second initial condition above equation determines

$$y(\pi/2) = -1 = (1/5)\{\cos \pi + 4\cos(3\pi/2)\} + (a/3)\sin(3\pi/2)$$

$$= -(1/5)\{1 + 4\cos(\pi/2)\} - (a/3)\sin(\pi/2) = -1/5 - a/3 \;\Rightarrow\; a/3 = 4/5.$$

Therefore, Eq. (2.10) reduces to

$$y(t) = (\cos 2t + 4\cos 3t + 4\sin 3t)/5. \;//$$

Example 2.3. Solve the differential equation

$$y''(t) - 3y'(t) + 2\, y(t) = 4\, t + 12\, e^{-t},$$

by using the Laplace transform methods when

$$y(0) = 0, \; y'(0) = 7. \tag{2.11}$$

Solution. Taking the Laplace transform of the differential equation, we get

$$£\, y''(t) - 3\, £\, y'(t) + 2\, £\, y(t) = 4\, £\,(t) + 12\, £\,(e^{-t}),$$

or, for Eqs. (2.3), (2.4), (6.2.2) and (6.2.4)

$$\{s^2\, £\, y(t) - s\, y(0) - y'(0)\} - 3\{s\, £\, y(t) - y(0)\} + 2\, £\, y(t)$$

$$= 4 / s^2 + 12 / (s + 1),$$

or, for Eq. (2.11)

$$(s^2 - 3s + 2)\, £\, y(t) \equiv (s - 1)(s - 2)\, £\, y(t) = 4/s^2 + 12/(s + 1) + 7$$

$\Rightarrow$

$$£\, y(t) = \{ 4/s^2 + 12 / (s + 1) + 7 \} / (s - 1)(s - 2). \tag{2.12}$$

Breaking into partial fractions:

$$4 / s^2 (s - 1)(s - 2) = a/s + b/s^2 + c / (s - 1) + d / (s - 2) \tag{2.13}$$

$\Rightarrow$

$$4 = a\, s\,(s - 1)(s - 2) + b\,(s - 1)(s - 2) + c\, s^2 (s - 2) + d.\, s^2 (s - 1).$$

Choosing $s = 0, 1, 2$ and -1 (say), we evaluate the coefficients:

$$b = 2,\; c = -4,\; d = 1, \text{ and } 4 = -6a + 12 + 12 - 2 \quad \Rightarrow \quad a = 3.$$

Therefore, Eq. (2.13) becomes

$$4/s^2 (s - 1)(s - 2) = 3/s + 2/s^2 - 4/(s - 1) + 1/(s - 2).$$

Similarly,

$$12 / (s + 1)(s - 1)(s - 2) = 2 / (s + 1) - 6 / (s - 1) + 4 / (s - 2),$$

and

$$7 / (s - 1)(s - 2) = 7 \{-1 / (s - 1) + 1 / (s - 2)\}.$$

So, Eq. (2.12) becomes

$$£\, y(t) = \{3/s + 2/s^2 - 4/(s - 1) + 1/(s - 2)\}$$

$$= 3/s + 2/s^2 - 17\,/\,(s-1) + 12\,/\,(s-2) + 2\,/\,(s+1).$$

Taking its inverse Laplace transform, we obtain the solution of the given differential equation:

$$y\,(t) = 3£^{-1}\,(1/s) + 2\,£^{-1}\,(1/s^2) - 17\,£^{-1}\{1/(s-1)\} + 12\,£^{-1}\,\{1/(s-2)\}$$

$$+\,2\,£^{-1}\,\{1/\,(s+1)\} = 3 + 2t - 17e^{\,t} + 12\,e^{\,2\,t} + 2e^{-t}, \quad \text{by Chapter 6. //}$$

Example 2.4. Using the Laplace transform methods solve the differential equation

$$(D^2 - 3\,D + 2)\,y\,(t) = 1 - e^{-2\,t}, \quad D \equiv d\,/dt,$$

when

$$y\,(0) = 1, \;\; y\,'\,(0) = 0. \tag{2.14}$$

Solution. Taking the Laplace transform of the differential equation, we get

$$£\,y''\,(t) - 3\,£\,y'\,(t) + 2\,£\,y\,(t) = £\,(1) - £\,(e^{-2t}),$$

or, for Eqs. (2.3), (2.4), (6.2.1) and (6.2.4)

$$\{s^2\,£\,y\,(t) - s\,y\,(0) - y'\,(0)\} - 3\{s\,£\,y\,(t) - y\,(0)\} + 2\,£\,y\,(t)$$

$$= 1/s - 1\,/\,(s+2),$$

or, for Eq. (2.14)

$$(s^2 - 3s + 2)\,£\,y\,(t) \equiv (s-1)(s-2)\,£\,y\,(t) = 1/s - 1/(s+2) + (s-3)$$

$\Rightarrow$

$$£\,y\,(t) = 1/s\,(s-1)(s-2) - 1/(s-1)(s-2)(s+2) + (s-3)/(s-1)(s-2)$$

$$= \{1/2s - 1/(s-1) + 1/2(s-2)\} + \{1/3(s-1) - 1/4(s-2) - 1/12(s+2)\}$$

$$+ \{2/\,(s-1) - 1/\,(s-2)\} = 1/2s + 4/3(s-1) - 3/4\,(s-2) - 1/12(s+2).$$

Taking its inverse Laplace transform, we obtain the solution of the given differential equation:

$$y\,(t) = (1/2)\,£^{-1}\,(1/s) + (4/3)\,£^{-1}\,\{1/\,(s-1)\} - (3/4)\,£^{-1}\,\{1/(s-2)\}$$

$$-\,(1/12)\,£^{-1}\,\{1/(s+2)\} = 1/2 + (4/3)\,e^{\,t} - (3/4)\,e^{2t} - (1/12)\,e^{-2\,t},$$

by Chapter 6. //

Example 2.5. Solve the following simultaneous differential equations by the Laplace transform methods:

$$dx/dt + 5x - 2y = t, \text{ and } dy/dt + 2x + y = 0, \; x(0) = y(0) = 0. \quad (2.15)$$

Solution. Taking Laplace transform of these equations and putting from Eqs. (6.2.2), (6.5.1) and (2.15), we get

$$\{s\,x(s) - x(0)\} + 5x(s) - 2y(s) = 1/s^2 \Rightarrow (s+5)\,x(s) - 2y(s) = 1/s^2$$

and

$$\{s\,y(s) - y(0)\} + 2x(s) + y(s) = 0 \Rightarrow 2x(s) + (s+1)\,y(s) = 0.$$

Solving these simultaneous equations for $x(s)$ and $y(s)$:

$$x(s) \equiv \pounds(x) = (s+1)/s^2(s+3)^2 = \{1/s + 3/s^2 - 1/(s+3) - 6/(s+3)^2\}/27,$$

and

$$y(s) \equiv \pounds(y) = -2/s^2(s+3)^2 = \{4/s - 6/s^2 - 4/(s+3) - 6/(s+3)^2\}/27.$$

Taking the inverse Laplace transform of above equation and making substitutions from Chapter 6, we obtain the solution:

$$x = (1 + 3t - e^{-3t} - 6t.\,e^{-3t})\,/\,27,$$

and

$$y = (4 - 6t - 4e^{-3t} - 6t.\,e^{-3t})\,/\,27. \;//$$

§ 3. Solution of Ordinary Differential Equations with Variable Coefficients

The method is demonstrated by means of the following examples.

Example 3.1. Using Laplace transform techniques solve the differential equation

$$\{t\,D^2 + (1-2t)\,D - 2\}\,y(t) = 0, \quad Dy \equiv dy/dt \equiv y'(t), \quad (3.1)$$

when

$$y(0) = 1, \; y'(0) = 2. \quad (3.2)$$

Solution. Taking the Laplace transform of the differential equation, we get

$$£\,(t\,y'') + £\,y' - 2\,£\,(t\,y') - 2\,£\,y = 0. \tag{3.3}$$

Applying Eq. (6.7.1a) and putting from Eqs. (2.3) and (2.4), we derive

$$£\,(t\,y'') = -\,(d/ds)\,£\,(y'') = -\,(d/ds)\{s^2\,£\,y\,(t) - s\,y\,(0) - y'\,(0)\}, \tag{3.4}$$

and

$$£\,(t\,y') = -\,(d/ds)\,£\,(y') = -\,(d/ds)\{s\,£\,y\,(t) - y\,(0)\}, \tag{3.5}$$

which reduce Eq. (3.3) to the form

$$-\,(d/ds)\,\{s^2\,£\,y\,(t) - s\,y\,(0) - y'\,(0)\} + \{s\,£\,y\,(t) - y\,(0)\}$$

$$+\,2\,(d/ds)\,\{s\,£\,y\,(t) - y\,(0)\} - 2\,£\,y\,(t) = 0.$$

Writing $£\,y\,(t) = y\,(s) = z$ (say), and putting from Eq. (3.2), above equation simplifies to

$$-\,(d/ds)\,(s^2\,z - s - 2) + (s\,z - 1) + 2(d/ds)\,(s\,z - 1) - 2z = 0,$$

or,

$$-\,2s\,z - s^2\,(dz/ds) + 1 + s\,z - 1 + 2z + 2s\,(dz/ds) - 2z = 0,$$

or, on division by $-\,s$,

$$(s-2)\,(dz\,/\,ds) + z = 0 \quad \Rightarrow \quad dz\,/\,z + ds\,/\,(s-2) = 0.$$

Integrating it term-wise, we get

$$\ln z + \ln\,(s-2) = \ln c \quad \Rightarrow \quad z = y\,(s) = c\,/\,(s-2).$$

Taking its inverse Laplace transform, we get

$$£^{-1}\,\{\,y\,(s)\} = c\,£^{-1}\,\{1\,/\,(s-2)\}, \text{ or } \quad y\,(t) = c\,e^{2t}.$$

Applying the first initial condition, we evaluate the constant $c = y\,(0) = 1$. Hence, the solution is $y\,(t) = e^{2t}$. //

Example 3.2. Using Laplace transform techniques solve the differential equation

$$\{t\,D^2 + t\,D - 1)\,y\,(t) = 0,\; Dy \equiv dy/dt \equiv y'\,(t), \tag{3.6}$$

if

$$y(0) = 0, \qquad y'(0) = 1. \tag{3.7}$$

Solution. Taking the Laplace transform of the differential equation, we get

$$£(t\,y'') + £(t\,y') - £\,y = 0, \tag{3.8}$$

or, for Eqs. (3.4) and (3.5)

$$(d/ds)\{s^2\,£\,y(t) - s\,y(0) - y'(0)\} + (d/ds)\{s\,£\,y(t) - y(0)\} + £\,y(t) = 0.$$

Writing $£\,y(t) = y(s) = z$ (say), and putting from Eq. (3.7), above equation simplifies to

$$(d/ds)(s^2 z - 1) + (d/ds)(s\,z) + z = 0,$$

or,

$$2s\,z + s^2(dz/ds) + z + s\,(dz/ds) + z = 0,$$

or

$$(s^2 + s)(dz/ds) + 2(s+1)z = 0 \Rightarrow dz/z + 2\,ds/s = 0.$$

Integrating it term-wise, we get

$$\ln z + 2\ln s = \ln c \Rightarrow z = y(s) = c/s^2.$$

Taking its inverse Laplace transform, we get

$$£^{-1}\{y(s)\} = c\,£^{-1}(1/s^2), \quad \text{or} \quad y(t) = c\,t. \tag{3.9}$$

Differentiating (3.9) w.r.t. t and applying the second initial condition, we evaluate the constant $c = y'(0) = 1$. Hence, the solution is $y(t) = t$. //

§ 4. Problem Set

4.1. Solve the following differential equations by the Laplace transform methods:

(i) $dx/dt + x = \sin at,\ x(0) = 2;$

(ii) $y'' + y' - 2y = t,\ y(0) = 1,\ y'(0) = 0;$

(iii) $y'' + 4y' + 3y = e^{-t}, y(0) = y'(0) = 1;$

(iv) $d^2y/dt^2 + 2(dy/dt) - 3y = \sin t,\ y = dy/dt = 0$ when $t = 0;$

(v) $(D^2 + a^2)\, y(t) = \cos at,\ t > 0,\ y = Dy = 0$ at $t = 0$;

(vi) $(D^2 - 1)\, x(t) = a.\cosh t,\ x(0) = Dx(0) = 0$;

(vii) $y'' + 2y' + 5y = e^{t}.\sin t,\ y(0) = 0,\ y'(0) = 1$;

(viii) $(D^3 + 2D^2 - D - 2)\, y(t) = 0,\ y = 1,\ Dy = D^2 y = 2$ at $t = 0$;

(ix) $(D^3 - 3D^2 + 3D - 1)\, y(t) = t^2 e^{2t},\ y = 1,\ Dy = 0,\ D^2 y = -2$ at $t = 0$;

(x) $(D^4 - k^4)\, y(t) = 0,\ y = 1,\ Dy = D^2 y = D^3 y = 0$ at $t = 0$.

[**Hint: (i)** The Laplace transform of the differential equation, for Eqs. (6.5.1), (6.2.5b) and the initial condition, yields

$$(s + 1)\, £\, x(t) = 2 + a/(s^2 + a^2) \Rightarrow £\, x(t) = 2/(s + 1) + a/(s + 1)(s^2 + a^2)$$

$$\Rightarrow \qquad x(t) = 2\, £^{-1}\{1/(s + 1)\}$$

$$+ \{a/(a^2 + 1)\}\, £^{-1}\{1/(s + 1) - s/(s^2 + a^2) + 1/(s^2 + a^2)\}$$

$$= \{2 + a/(a^2 + 1)\}\, e^{-t} + (-a.\cos at + \sin at)/(a^2 + 1),$$

by Eqs. (6.2.4) and (6.2.5).

(ii) The Laplace transform of the differential equation, for Eqs. (6.2.2), (6.5.1), (6.5.2) and the initial condition, yields

$$(s^2 + s - 2)\, £\, y(t) = (s + 1) + 1/s^2$$

$$\Rightarrow \qquad £\, y(t) = (s + 1)/(s - 1)(s + 2) + 1/s^2(s - 1)(s + 2)$$

$$= 1/(s - 1) + 1/4(s + 2) - 1/4s - 1/2s^2$$

$$\Rightarrow \qquad y(t) = £^{-1}\{1/(s - 1) + 1/4(s + 2) - 1/4s - 1/2s^2\}$$

$$= e^{t} + (1/4)(e^{-2t} - 1) - t/2.$$

(iii) The Laplace transform of the differential equation, for Eqs. (6.2.4), (6.5.1), (6.5.2) and the initial conditions, yields

$$(s^2 + 4s + 3)\,£\,y(t) = (s + 5) + 1/(s + 1)$$

$\Rightarrow$

$$£\,y(t) = (s + 5)/(s + 1)(s + 3) + 1/(s + 1)^2(s + 3)$$

$$= 7/4(s + 1) - 3/4(s + 3) + 1/2(s + 1)^2$$

$\Rightarrow$

$$y(t) = (1/4)\,£^{-1}\{7/(s + 1) - 3/(s + 3) - 2\,(d/ds)\,(s + 1)^{-1}\}$$

$$= (7e^{-t} - 3e^{-3t} + 2t.\,e^{-t})/4 = \{(2t + 7)\,e^{-t} - 3e^{-3t}\}/4,$$

by Eqs. (6.2.4) and (7.4.2).

(iv) The Laplace transform of the differential equation, for Eqs. (6.2.5b), (6.5.1), (6.5.2) and the initial conditions, yields

$$(s^2 + 2s - 3)\,£\,y(t) = 1/(s^2 + 1)$$

$\Rightarrow$

$$£\,y(t) = 1/(s - 1)(s + 3)(s^2 + 1)$$

$$= 1/8(s - 1) - 1/40(s + 3) - s/10(s^2 + 1) - 1/5(s^2 + 1)$$

$\Rightarrow$

$$y(t) = e^{t}/8 - e^{-3t}/40 - (\cos t)/10 - (\sin t)/5,$$

by Eqs. (6.2.4) and (6.2.5).

(v) The Laplace transform of the differential equation, for Eqs. (6.2.5a), (6.5.2) and the initial conditions, yields

$$(s^2 + a^2)\,£\,y(t) = s/(s^2 + a^2) \quad \Rightarrow \quad £\,y(t) = s/(s^2 + a^2)^2$$

$\Rightarrow$

$$y(t) = £^{-1}\{s/(s^2 + a^2)^2\} = (t/2a)\sin at, \qquad \text{by Eq. (7.3.2).}$$

(vi) The Laplace transform of the differential equation, for Eqs. (6.2.6a), (6.5.2) and the initial conditions, yields

$$(s^2 - 1)\,£\,x(t) = as/(s^2 - 1) \quad \Rightarrow \quad £\,x(t) = as/(s^2 - 1)^2$$

$$= -(a/2)\,(d/ds)\,(s^2 - 1)^{-1}$$

$\Rightarrow$

$$x(t) = -(a/2).\,£^{-1}\{(d/ds)\,(s^2 - 1)^{-1}\}$$

$$= (at/2)\,£^{-1}\{1/(s^2-1)\}, \qquad \text{by Eq. (7.4.2)}$$

$$= (at/2)\sinh t, \qquad \text{by Eq. (6.2.6b).}$$

(vii) The Laplace transform of the differential equation, for Eqs. (6.3.2), (6.5.1), (6.5.2) and the initial conditions, yields

$$(s^2+2s+5)\,£\,y(t) = 1 + f(s-1), \quad \text{where} \quad f(s) = £\sin t = 1/(s^2+1)$$

$$\Rightarrow \qquad £\,y(t) = 1/(s^2+2s+5) + 1/(s^2+2s+5)(s^2-2s+2). \qquad (4.1)$$

The second fraction splits into partial fractions:

$$(1/65)\,\{\,(4s+5)/(s^2+2s+5) + (11-4s)/(s^2-2s+2)\,\}$$

$$= [4\,(s+1)/\{(s+1)^2+4\} + 1/\{(s+1)^2+4\} - 4\,(s-1)/\{(s-1)^2+1\}$$

$$+\,7/\{(s-1)^2+1\}]/65,$$

and its inverse Laplace transform, for Eq. (7.2.2), becomes

$$(e^{-t}\,[\,4\,£^{-1}\{s/(s^2+4)\} + £^{-1}\{1/(s^2+4)\}\,]$$

$$+\,e^{t}\,[\,-4\,£^{-1}\{s/(s^2+1)\} + 7\,£^{-1}\{1/(s^2+1)\}\,]\,)/65$$

$$= [e^{-t}\{4\cos 2t + (\sin 2t)/2\} + e^{t}(-4\cos t + 7\sin t)]/65,$$

by Eq. (6.2.5). Similarly, the inverse Laplace transform of the first fraction in Eq. (4.1) becomes

$$£^{-1}[1/\{(s+1)^2+4\}] = e^{-t}£^{-1}\{1/(s^2+4)\} = (1/2)\,e^{-t}\sin 2t.$$

Hence, Eq. (4.1) yields

$$y(t) = (1/65)\{e^{-t}(4\cos 2t + 33\sin 2t) + e^{t}(-4\cos t + 7\sin t)\}.$$

(viii) The Laplace transform of the third order derivative of $y(t)$, by Eq. (6.5.3) is

$$£\,y'''(t) = s^3\,£\,y(t) - s^2\,y(0) - s\,y'(0) - y''(0). \qquad (4.2)$$

Therefore, the Laplace transform of the given differential equation, for

Eqs. (6.5.1), (6.5.2), (4.2) and the initial conditions, yields

$$(s^3 + 2s^2 - s - 2)\,£\,y\,(t) = s^2 + 4s + 5$$

$\Rightarrow$

$$£\,y\,(t) = (s^2 + 4s + 5) / (s - 1)(s + 1)(s + 2)$$

$$= (1/3)\ \{ 5/(s - 1) - 3/(s + 1) + 1/(s + 2) \}$$

$\Rightarrow$

$$y\,(t) = (1/3)\,£^{-1}\{ 5/(s - 1) - 3/(s + 1) + 1/(s + 2) \}$$

$$= (1/3)\,(5\,e^{\,t} - 3\,e^{-t} + e^{-2t}).$$

(ix) The Laplace transform of $t^2.e^{2t}$ may be found by Eq. (6.7.1):

$$£\,(t^2 e^{2t}) = (d^2/ds^2)\,£\,(e^{2t}) = (d^2/ds^2)\,(s - 2)^{-1} = 2/(s - 2)^3. \quad (4.3)$$

Hence, the Laplace transform of the given differential equation, for Eqs. (6.5.1), (6.5.2), (4.3) and the initial conditions, yields

$$(s^3 - 3s^2 + 3s - 1)\,£\,y\,(t) = s^2 - 3s + 1 + 2/(s - 2)^3$$

$\Rightarrow$

$$£\,y\,(t) = (s^2 - 3s + 1) / (s - 1)^3 + 2 / (s - 1)^3 (s - 2)^3. \qquad (4.4)$$

The first term splits as $1/(s - 1) - 1/(s - 1)^2 - 1/(s - 1)^3$, whereas the second term breaks into the following partial fractions:

$$12/(s - 2) - 6/(s - 2)^2 + 2/(s - 2)^3 - 12/(s - 1) - 6/(s - 1)^2 - 2/(s - 1)^3.$$

Hence, Eq. (4.4) yields

$$y\,(t) = £^{-1}\{12/(s - 2) + 6\,(d/ds)\,(s - 2)^{-1} + (d^2/ds^2)\,(s - 2)^{-1}$$

$$- 11/(s - 1) + 7\,(d/ds)\,(s - 1)^{-1} - (3/2)\,(d^2/ds^2)\,(s - 1)^{-1}\}$$

$$= 12\,e^{2t} - 6t\,e^{2t} + t^2 e^{2t} - 11\,e^{\,t} - 7\,t\,e^t - (3/2)\,t^2 e^t,$$

by Eqs. (6.2.4), (7.4.2) and (4.3), it becomes

$$= (12 - 6t + t^2)\,e^{2t} - (11 + 7t + 3t^2/2)\,e^{\,t}.$$

(x) The Laplace transform of the given differential equation, for Eq. (6.5.3) and the initial conditions, yields

$$(s^4 - k^4)\, £\, y(t) = s^3$$

$\Rightarrow$

$$£\, y(t) = s^3 / (s - k)(s + k)(s^2 + k^2)$$

$$= (1/4)\,\{1/(s - k) + 1/(s + k) + 2\, s/(s^2 + k^2)\}$$

$\Rightarrow$

$$£\, y(t) = (1/4)\{e^{kt} + e^{-kt} + 2\cos(k\, t)\}].$$

4.2. Solve the following simultaneous differential equations by the Laplace transform methods:

(i) $D^2 x + 3x - 2y = 0, \qquad D^2 x + D^2 y - 3x + 5y = 0,$

$$(x = y = 0,\ Dx = 3,\ Dy = 2 \text{ at } t = 0);$$

(ii) $(D^2 - 1)\, x + 5D\, y = t, \qquad -2Dx + (D^2 - 4)\, y = -2,$

$$(x = y = Dx = Dy = 0 \text{ at } t = 0);$$

(iii) $dx/dt - y = e^t$ and $dy/dt + x = \sin t$, $x(0) = 1$, $y(0) = 0$;

(iv) $3dx/dt + dy/dt + 2x = 1$, $dx/dt + 4dy/dt + 3y = 0$,

$$x(0) = 3,\ y(0) = 0;$$

(v) $dx/dt + dy/dt = t, \qquad d^2x/dt^2 - y = e^{-t},$

$$x = 3,\ y = 0,\ dx/dt = -2 \text{ at } t = 0;$$

(vi) $dx/dt = 2x - 3y$, $dy/dt = y - 2x$, $x(0) = 8$, $y(0) = 3$;

(vii) $dx/dt - dy/dt - 2x - y = 6\, e^{3t}, \qquad 2dx/dt + dy/dt - 3x - 3y = 6\, e^{3t},$

$$x(0) = 3,\ y(0) = 0.$$

[**Hint: (i)** The Laplace transform of the given differential equations, for Eqs. (6.5.1), (6.5.2) and the initial conditions, yields

$$(s^2 + 3)\, £\, x - 2\, £\, y = 3 \quad \text{and} \quad (s^2 - 3)\, £\, x + (s^2 + 5)\, £\, y = 5.$$

Solving these simultaneous equations for $£\, x$ and $£\, y$, we derive

$$£\, x = (3s^2 + 25)/(s^4 + 10s^2 + 9) = (3s^2 + 25)/(s^2 + 1)(s^2 + 9)$$

$$= (1/4)\ \{11\ /\ (s^2 + 1) + 1\ /\ (s^2 + 9)$$

$\Rightarrow$

$$x\,(t) = (11/4).\ £^{-1}\{1/\ (s^2 + 1)\} + (1/12).\ £^{-1}\{3\ /\ (s^2 + 9)\}$$

$$= (11/4).\ \sin t + (1/12).\ \sin 3t,$$

and

$$£\,y = 2(s^2 + 12)\ /\ (s^2 + 1)\ (s^2 + 9) = (1/4)\{11/\ (s^2 + 1) - 3/\ (s^2 + 9)\}$$

$\Rightarrow$

$$y\,(t) = (11/4).\ £^{-1}\{1/\ (s^2 + 1)\} - (1/4).\ £^{-1}\{3/\ (s^2 + 9)\}$$

$$= (11/4)\ \sin t - (1/4)\ \sin 3t.$$

(ii) The Laplace transform of the given differential equations, for Eqs. (6.2.1), (6.2.2), (6.5.1), (6.5.2) and the initial conditions, yields

$$(s^2 - 1)\ £\,x + 5s\ £\,y = 1/s^2, \qquad \text{and} \qquad 2s\ £\,x + (4 - s^2)\ £\,y = 2/s.$$

Solving these simultaneous equations for $£\,x$ and $£\,y$, we derive

$$£\,x = (11s^2 - 4)\ /\ s^2(s^2 + 1)\ (s^2 + 4) = -1/\ s^2 + 5/\ (s^2 + 1) - 4/\ (s^2 + 4)$$

$\Rightarrow$

$$x\,(t) = -£^{-1}\,(1/s^2) + 5.\ £^{-1}\{1/\ (s^2 + 1)\} - 2.\ £^{-1}\{2\ /\ (s^2 + 4)\}$$

$$= -\ t + 5.\ \sin t - 2.\ \sin 2t,$$

and

$$£\,y = (4 - 2s^2)\ /s\ (s^2 + 1)\ (s^2 + 4) = 1/s - 2s\ /\ (s^2 + 1) + s\ /\ (s^2 + 4)$$

$\Rightarrow$

$$y\,(t) = £^{-1}\{1/s - 2s\ /\ (s^2 + 1) + s\ /\ (s^2 + 4)\} = 1 - 2\cos t + \cos 2t,$$

by Eqs. (6.2.1) and (6.2.5a).

(iii) The Laplace transform of the given differential equations, for Eqs. (6.2.4), (6.2.5b), (6.5.1) and the initial conditions, yields

$$s.\ £\,x - £\,y = s\ /\ (s - 1), \qquad \text{and} \qquad £\,x + s.\ £\,y = 1/\ (s^2 + 1).$$

Solving these simultaneous equations for $£\,x$ and $£\,y$, we derive

$$£\,x = s^2/\ (s - 1)\ (s^2 + 1) + 1\ /\ (s^2 + 1)^2$$

$$= (1/2)\ \{1\ /\ (s - 1) + s\ /\ (s^2 + 1) + 1\ /\ (s^2 + 1) + 1\ /\ (s^2 + 1)^2$$

$\Rightarrow$

$$x(t) = (1/2)\,£^{-1}\{1/(s-1) + s/(s^2+1) + 1/(s^2+1) + 1/(s^2+1)^2\}$$

$$= (1/2)\,(e^{t} + \cos t + 2\sin t - t.\cos t),$$

by Eqs. (6.2.4), (6.2.5), (7.6.2); and

$$£\,y = s/(s^2+1)^2 - s/(s-1)(s^2+1)$$

$$= s/(s^2+1)^2 + (1/2).\,\{-1/(s-1) + s/(s^2+1) - 1/(s^2+1)\}$$

$\Rightarrow$

$$y(t) = £^{-1}[s/(s^2+1)^2 + \{-1/(s-1) + s/(s^2+1) - 1/(s^2+1)\}/2]$$

$= (1/2)\,(t.\sin t - e^{t} + \cos t - \sin t)$, by Eqs. (6.2.4), (6.2.5) and (7.3.2).

(iv) The Laplace transform of the given differential equations, for Eqs. (6.2.1), (6.5.1) and the initial conditions, yields

$$(3s+2)\,£\,x + s\,£\,y = (9s+1)/s, \text{ and } \qquad s\,£\,x + (4s+3)\,£\,y = 3.$$

Solving these simultaneous equations for $£\,x$ and $£\,y$, we derive

$$£\,x = (33s^2 + 31s + 3)/s\,(11s+6)\,(s+1)$$

$$= 1/2s + 1/(s+1) + 3/2.\,(s + 6/11)$$

$\Rightarrow$

$$x(t) = (1/2)\,[£^{-1}(1/s) + 2£^{-1}\{1/(s+1)\} + 3£^{-1}\{1/(s+6/11)\}]$$

$= (1/2)\,(1 + 2e^{-t} + 3\,e^{-6t/11})$, by Eqs. (6.2.1) and (6.2.4);

and

$$£\,y = 5/(s+1)\,(11s+6) = -1/(s+1) + 11/(11s+6)$$

$\Rightarrow$

$y(t) = £^{-1}\{-1/(s+1) + 1/(s+6/11)\} = -e^{-t} + e^{-6t/11}$, by Eq. (6.2.4).

(v) The Laplace transform of the given differential equations, for Eqs. (6.2.2), (6.2.4), (6.5.1), (6.5.2) and the initial conditions, yields

$$s.\,£\,x + s.\,£\,y = 3 + 1/s^2, \text{ and } \quad s^2.\,£\,x - £\,y = 3s - 2 + 1/(s+1).$$

Solving these simultaneous equations for $£\,x$ and $£\,y$, we derive

$$£\,x = 3/s\,(s^2+1) + 1/s^3\,(s^2+1) + 3s/(s^2+1) - 2/(s^2+1)$$

$$+ 1 / (s + 1)(s^2 + 1)$$

$$= 2/s + (1/2)\{1/(s+1) - 3/(s^2+1) + s/(s^2+1) + 2/s^3\}$$

$\Rightarrow$

$$x(t) = 2.\ £^{-1}(1/s) + (1/2)\ £^{-1}\{1/(s+1) - 3/(s^2+1) + s/(s^2+1) + 2/s^3\}$$

$$= 2 + (1/2)(e^{-t} - 3.\sin t + \cos t + t^2),$$

by Eqs. (6.2.1), (6.2.3) - (6.2.5); and

$$£\,y = 2/(s^2+1) + 1/s(s^2+1) - 1/(s+1)(s^2+1)$$

$$= 1/s - (1/2).\ \{1/(s+1) - 3/(s^2+1) + s/(s^2+1)\}$$

$\Rightarrow$

$$y(t) = £^{-1}(1/s) - (1/2)\ £^{-1}\{1/(s+1) - 3/(s^2+1) + s/(s^2+1)\}/2$$

$= 1 - (1/2)(e^{-t} - 3\sin t + \cos t)$, by Eqs. (6.2.1), (6.2.4) and (6.2.5).

(vi) The Laplace transform of the given differential equations, for Eq. (6.5.1) and the initial conditions, yields

$$(s-2)\,£\,x + 3\,£\,y = 8, \qquad \text{and} \qquad 2\,£\,x + (s-1)\,£\,y = 3.$$

Solving these simultaneous equations for $£\,x$ and $£\,y$, we derive

$$£\,x = (8s - 17)/(s+1)(s-4) = 5/(s+1) + 3/(s-4)$$

$\Rightarrow$

$$x(t) = £^{-1}\{5/(s+1) + 3/(s-4)\} = 5e^{-t} + 3e^{4t}, \text{ by Eq. (6.2.4)}$$

and

$$£\,y = (3s - 22)/(s+1)(s-4) = 5/(s+1) - 2/(s-4)$$

$\Rightarrow$

$$y(t) = £^{-1}\{5/(s+1) - 2/(s-4)\} = 5e^{-t} - 2e^{4t}, \text{ again by Eq. (6.2.4).}$$

(vii) The Laplace transform of the given differential equations, for Eqs. (6.2.4), (6.5.1) and the initial conditions, yields

$$(s-2)\,£\,x - (s+1)\,£\,y = 3 + 6/(s-3),$$

and

$$(2s-3)\,£\,x + (s-3)\,£\,y = 6 + 6/(s-3).$$

Solving these simultaneous equations for $£\,x$ and $£\,y$, we derive

$$£\,x = (3s^2 - 6s - 1)/(s - 3)(s - 1)^2 = 1/(s - 1) + 2/(s - 3) + 2/(s - 1)^2$$

$\Rightarrow$

$$x\,(t) = £^{-1}\,\{1/(s - 1) + 2/(s - 3) - 2(d/ds)(s - 1)^{-1}\} = e^t + 2e^{3t} + 2t.\,e^t$$

by Eqs. (6.2.4) and (7.4.2); and

$$£\,y = (5 - 3s)\,/\,(s - 3)(s - 1)^2 = 1/\,(s - 1) - 1/\,(s - 3) - 1/\,(s - 1)^2$$

$\Rightarrow$

$$y\,(t) = £^{-1}\,\{1/\,(s - 1) - 1/\,(s - 3) + (d/ds)\,(s - 1)^{-1}\} = e^{\,t} - e^{\,3t} - t.\,e^{\,t},$$

by Eqs. (6.2.4) and (7.4.2).]

4.3. The coordinates of a particle describing a plane curve at any instant t are given by

$$dy/dt + 2x = \sin 2t, \qquad\qquad dx/dt - 2y = \cos 2t,\ \ t > 0.$$

If the particle started from a location (1, 0), show by techniques of Laplace transform that its locus is

$$4x^2 + 4xy + 5y^2 = 4.$$

[**Hint:** The Laplace transform of the given differential equations, for (6.2.5), (6.5.1) and the initial conditions, yields

$$2£\,x + s\,£\,y = 2\,/\,(s^2 + 4), \quad \text{and} \quad s\,£\,x - 2\,£\,y = 1 + s\,/\,(s^2 + 4).$$

Solving these simultaneous equations for $£\,x$ and $£\,y$, we derive

$$£\,x = (s + 1)\,/\,(s^2 + 4)$$

$\Rightarrow$

$$x\,(t) = £^{-1}\,\{s\,/\,(s^2 + 4) + 1\,/\,(s^2 + 4)\} = \cos 2t + (\sin 2t)\,/2,$$

and

$$£\,y = -2\,/\,(s^2 + 4) \ \Rightarrow\ y\,(t) = -\,£^{-1}\{2\,/\,(s^2 + 4)\} = -\sin 2t,$$

by Eq. (6.2.5). Eliminating sin $2t$ from above relations, we derive the locus:

$$\{x - (\sin 2t)\,/\,2\}^2 = \cos^2 2t = 1 - \sin^2 2t = 1 - y^2,$$

or,

$$(x + y/2)^2 \equiv x^2 + x\,y + y^2/4 = 1 - y^2, \ \text{ or } \ 4x^2 + 4xy + 5y^2 = 4.]$$

4.4. The currents i_1 , i_2 in are given by the differential equations

$$d\,i_1 / dt - \omega\, i_2 = a.\cos pt, \quad d\,i_2 / dt + \omega\, i_1 = a.\sin pt$$

Find the currents by Laplace transform methods if $i_1 = i_2 = 0$ at $t = 0$.

[**Hint:** The Laplace transform of the given differential equations, for Eqs. (6.2.5), (6.5.1) and the initial conditions, yields

$$s\,£\,i_1 - \omega\,£\,i_2 = as / (s^2 + p^2), \quad \text{and} \quad \omega\,£\,i_1 + s\,£\,i_2 = ap / (s^2 + p^2).$$

Solving these simultaneous equations for $£\,i_1$ and $£\,i_2$, we derive

$$£\,i_1 = a\,(s^2 + \omega p) / (s^2 + \omega^2)\,(s^2 + p^2)$$

$$= \{a / (p + \omega)\}\,\{\omega / (s^2 + \omega^2) + p / (s^2 + p^2)\}$$

$\Rightarrow$

$$i_1\,(t) = \{a / (p + \omega)\}.\,£^{-1}\{\,\omega / (s^2 + \omega^2) + p / (s^2 + p^2)\,\}$$

$$= \{a / (p + \omega)\}\,(\sin \omega t + \sin pt), \qquad \text{by Eq. (6.2.5b)}$$

and

$$£\,i_2 = a\,(p - \omega)\,s / (s^2 + \omega^2)\,(s^2 + p^2)$$

$$= \{a / (p + \omega)\}\,\{s / (s^2 + \omega^2) - s / (s^2 + p^2)\}$$

$\Rightarrow$

$$i_2\,(t) = \{a / (p + \omega)\}.\,£^{-1}\{\,s / (s^2 + \omega^2) - s / (s^2 + p^2)\}$$

$$= \{a / (p + \omega)\}\,(\cos \omega t - \cos pt), \qquad \text{by Eq. (6.2.5a).}$$]

CHAPTER 9

FOURIER TRANSFORMS OF FUNCTIONS

§ 1. Introduction

The Fourier transform of the function f is traditionally denoted by adding a circumflex: $\hat{f}$. There are several common conventions for defining the Fourier transform of an integrable function f: R → C.

Here we use the following definition:

$$\hat{f}(t) = \int_{-\infty}^{\infty} f(x) \,.\, e^{-2i\pi xt} .\, dt$$

for any real number t. When the independent variable x represents *time*, the transform variable t represents frequency (e.g. if time is measured in seconds, then the frequency is in *hertz*). Under suitable conditions, f is determined by $\hat{f}$ via the **inverse transform**:

$$f(x) = \int_{-\infty}^{\infty} \hat{f}(t) \cdot e^{2i\pi xt} .\, dt$$

for any real number x.

The reason for the negative sign convention in the definition of $\hat{f}(t)$ is that the integral produces the amplitude and phase of the function $f(x).\, e^{-2i\pi xt}$ at frequency zero (0), which is identical to the amplitude and phase of the function $f(x)$ at frequency t, which is what $\hat{f}(t)$ is supposed to represent.

Note 1.1. There is a close connection between the definition of Fourier series and the Fourier transform for functions f that are zero outside an interval. For such a function, we can calculate its Fourier series on any interval that includes the points where f is not identically zero. The Fourier transform is also defined for such a function. As we increase the length of the interval on which we calculate the Fourier series, then the Fourier series coefficients begin to look like the Fourier transform and the sum of the Fourier series of f begins to look like the inverse Fourier transform.

§ 2. Properties of the Fourier Transform

Here we consider $f(x)$, $g(x)$ and $h(x)$ as *integrable functions*: Lebesgue–measurable on the real line satisfying:

$$\int_{-\infty}^{\infty} |f(x)|\, dx < \infty.$$

We denote the Fourier transforms of these functions by $\hat{f}(t)$, $\hat{g}(t)$ and $\hat{h}(t)$ respectively.

The Fourier transform has the following basic properties:

(i) **Linearity:** For any complex numbers a and b, if

$$h(x) = a f(x) + b\, g(x), \text{ then } \hat{h}(t) = a \cdot \hat{f}(t) + b \cdot \hat{g}(t).$$

(ii) **Translation / time shifting:** For any real number x_o, if

$$h(x) = f(x - x_o), \text{ then } \hat{h}(t) = \exp(-2\pi i x_o t).\, \hat{f}(t).$$

(iii) **Modulation / frequency shifting:** For any real number t_o, if

$$h(x) = \exp(2\pi i x t_o).\, f(x), \text{ then } \hat{h}(t) = \hat{f}(t - t_o).$$

(iv) **Time scaling:** For a non–zero real number a, if

$$h(x) = f(ax), \text{ then } \hat{h}(t) = (1/|a|)\hat{f}(t/a).$$

The case $a = -1$ leads to the *time–reversal* property, which states: if

$$h(x) = f(-x), \text{ then } \hat{h}(t) = \hat{f}(-t).$$

§ 3. Fourier Series

Let $f(x)$ be some function of x defined over some interval, say $[-\pi, \pi]$ of the real line and it satisfies certain conditions. We know, by Taylor's theorem of differential calculus, the function can be expanded in infinite series of terms in polynomial form about a given point x_o in the interval:

$$f(x) = \sum_{n=0}^{\infty} a_n (x - x_o)^n. \tag{3.1}$$

Expansion of $f(x)$ in an infinite series whose terms need not be polynomials is also possible. One of the most useful such expansion is the following trigonometric series:

$$\begin{aligned} f(x) &= a_0 + (a_1 \cos x + a_2 \cos 2x + \ldots + a_n \cos nx + \ldots) \\ &\quad + (b_1 \sin x + b_2 \sin 2x + \ldots + b_n \sin nx + \ldots) \\ &= a_0 + \sum_{n=1}^{\infty} (a_n \cos nx + b_n \sin nx), \end{aligned} \tag{3.2}$$

so that each term in the series is term-wise integrable. Above series is called a *Fourier series* for the function $f(x)$ over the given interval $[-\pi, \pi]$. The coefficients $a_0, a_1, a_2, \ldots, a_n; b_1, b_2, \ldots, b_n$ are called the *Fourier coefficients.*

§ 4. Evaluation of Fourier Coefficients

4.1. Integrating the series in Eq. (3.2) over the interval $[-\pi, \pi]$, and applying the properties of definite integrals:

$$\int_{-\pi}^{\pi} \cos nx\, dx = 2\int_{0}^{\pi} \cos nx\, dx = (2/n)\left[\sin nx\right]_0^{\pi} = 0 \tag{4.1}$$

$$\int_{-\pi}^{\pi} \sin nx\, dx = 0, \text{ (} \sin x \text{ being an odd function of } x\text{)} \tag{4.2}$$

we evaluate the Fourier coefficients:

$$\int_{-\pi}^{\pi} f(x)\, dx = a_0 \int_{-\pi}^{\pi} 1\, dx = 2a_0 \int_{0}^{\pi} 1\, dx = 2\pi a_0$$

$$\Rightarrow \qquad a_0 = (1/2\pi)\int_{-\pi}^{\pi} f(x)\, dx. \tag{4.3}$$

4.2. Multiplying the series in Eq. (3.2) by $\cos mx$ and integrating the result so obtained over the interval $[-\pi, \pi]$, we get

$$\int_{-\pi}^{\pi} f(x).\cos mx.\, dx = a_0.\int_{-\pi}^{\pi} \cos mx.\, dx$$

$$+\sum_{n=1}^{\infty} a_n.\int_{-\pi}^{\pi} \cos mx. \cos nx. dx + \sum_{n=1}^{\infty} b_n \int_{-\pi}^{\pi} \cos mx. \sin nx\, dx. \quad (4.4)$$

The integrands cos mx. cos nx (respectively cos mx. sin nx) being even (resp. odd) functions, for Eqs. (4.1), (4.2) and the properties of definite integrals, Eq. (4.4) simplifies to

$$\int_{-\pi}^{\pi} f(x).\cos mx\, dx = \sum_{n=1}^{\infty} a_n \int_{0}^{\pi} \{\cos (m-n)\, x + \cos (m+n)\, x\, dx \quad (4.5)$$

$$= \sum_{n=1}^{\infty} a_n [\{\sin (m-n)x\}/(m-n) + \{\sin (m+n)x\}/(m+n)]_0^{\pi} = 0,$$

when $n \neq m$ (4.6)

On the other hand, when $n = m$, Eq. (4.5) determines

$$\int_{-\pi}^{\pi} f(x) \cos mx\, dx = a_m \int_{0}^{\pi} \{1 + \cos 2mx)\, dx$$

$$= a_m \left[x + (\sin 2mx)/2m\right]_0^{\pi} = \pi a_m$$

$$\Rightarrow \qquad a_m = (1/\pi) \int_{-\pi}^{\pi} f(x). \cos mx\, dx. \qquad (4.7)$$

4.3. Multiplying the series in Eq. (3.2) by sin mx and integrating the result so obtained over the interval $(-\pi, \pi)$, we also find

$$\int_{-\pi}^{\pi} f(x) \sin mx\, dx = a_0 \int_{-\pi}^{\pi} \sin mx + \sum_{n=1}^{\infty} a_n \int_{-\pi}^{\pi} \sin mx. \cos nx\, dx$$

$$+ b_n \int_{-\pi}^{\pi} \sin mx. \sin nx\, dx.$$

Again, for Eq. (4.2) and the properties of definite integrals, above equation simplifies to

$$\int_{-\pi}^{\pi} f(x).\sin mx\, dx = \sum_{n=1}^{\infty} b_n \int_{0}^{\pi} \{\cos (m-n)\, x - \cos (m+n)\, x\, dx$$

(4.8)

$$= \sum_{n=1}^{\infty} b_n\, [\{\sin (m-n)x\}/(m-n) - \{\sin (m+n)x\}/(m+n)]_0^{\pi} = 0,$$

when $n \neq m$. On the other hand, when $n = m$, Eq. (4.8) determines

$$\int_{-\pi}^{\pi} f(x) \sin mx \, dx = b_m \int_{0}^{\pi} (1 - \cos 2mx) \, dx$$

$$= b_m \left[x - (\sin 2mx)/2m\right]_0^{\pi} = \pi b_m$$

$\Rightarrow$

$$b_m = (1/\pi) \int_{-\pi}^{\pi} f(x). \sin mx \, dx, \tag{4.9}$$

Putting from Eqs. (4.3), (4.7) and (4.9), the Fourier series in Eq. (3.2) for the function $f(x)$ assumes the form:

$$f(x) = (1/2\pi)\int_{-\pi}^{\pi} f(x)\, dx + (1/\pi) \sum_{n=1}^{\infty} [\, \{ \int_{-\pi}^{\pi} f(x) \cos nx.\, dx \} \cos nx$$

$$+ \{ \int_{-\pi}^{\pi} f(x). \sin nx.\, dx \}. \sin nx \,] \tag{4.10}$$

$$= (1/2\pi)\int_{-\pi}^{\pi} f(t)\, dt + (1/\pi) \sum_{n=1}^{\infty} \int_{-\pi}^{\pi} f(t) \{ \cos nt.\, \cos nx$$

$$+ \sin nt.\, \sin nx \}\, dt$$

$$= (1/2\pi) \int_{-\pi}^{\pi} f(t)\, dt + (1/\pi) \sum_{n=1}^{\infty} \int_{-\pi}^{\pi} f(t). \cos \{n\,(t - x)\}\, dt. \tag{4.11}$$

4.4. Particular cases

(i) When $f(x)$ is an even function of x, i.e. $f(-x) = f(x)$, the Fourier coefficients reduce to

$$a_0 = (1/\pi)\int_{0}^{\pi} f(x)\, dx, \quad a_n = (2/\pi)\int_{0}^{\pi} f(x).\cos nx.\, dx, \quad b_n = 0, \tag{4.12}$$

for the odd character of the integrand in Eq. (4.9) causing vanishing of the integral. Consequently, the Fourier series given by Eq. (4.10), for such a function, reduces to

$$f(x) = a_0 + \sum_{n=1}^{\infty} a_n \cos nx = (1/\pi)\, [\int_{0}^{\pi} f(x)\, dx$$

$$+ 2\sum_{n=1}^{\infty} \{ \int_{0}^{\pi} f(x). \cos nx\, dx\} \cos nx] \tag{4.10a}$$

(ii) On the other hand, if $f(x)$ is an odd function of x, i.e. $f(-x) = -f(x)$, the Fourier coefficients reduce to

$$a_0 = 0, \quad a_n = 0, \quad b_n = (2/\pi)\int_0^{\pi} f(x).\sin nx\, dx \qquad (4.13)$$

again for the odd character of the integrand in Eq. (4.7) causing vanishing of the integral. Consequently, the Fourier series given by Eq. (4.10), for an odd function, reduces to

$$f(x) = \sum_{n=1}^{\infty} b_n \sin nx = (2/\pi)\sum_{n=1}^{\infty} \{\int_0^{\pi} f(x).\sin nx\, dx\}.\sin nx].$$

(4.10b)

Thus, we conclude that:

(i) The Fourier series for an *even* function consists of a constant term a_0 and the *cosine* series;

(ii) The Fourier series for an *odd* function consists of only a *sine* series.

Example 4.1. Find the Fourier series for the function $f(x) = x$ defined in the interval $[-\pi, \pi]$.

Solution. The function being odd, the Fourier coefficients can be evaluated by (4.13): $a_0 = 0$, $a_n = 0$, and

$$b_n = (2/\pi)\int_0^{\pi} x.\sin nx.\, dx = (2/n\pi)\{\left[-x.\cos nx\right]_0^{\pi} + \int_0^{\pi} \cos nx.\, dx\}$$

$$= (2/n\pi)\{-\pi.\cos n\pi + (1/n).\left[\sin nx\right]_0^{\pi}\}$$

$$= (2/n\pi)\{-\pi.\cos n\pi + (1/n)\sin n\pi\} = -(2/n).\cos n\pi = (2/n)(-1)^{n+1}.$$

Hence, the Fourier series in Eq. (3.2) reduces to

$$f(x) \equiv x = \sum_{n=1}^{\infty} (2/n)(-1)^{n+1}.\sin nx$$

$$= 2\{(\sin x)/1 - (\sin 2x)/2 + (\sin 3x)/3 - (\sin 4x)/4 + \ldots \infty\}.// \quad (4.14)$$

Example 4.2. Find a series of *sines* and *cosines* of multiples of x which shall represent the function $f(x) = x + x^2$ in the interval $[-\pi, \pi]$. Hence, show that

$$\pi^2 / 6 = 1 + 1/2^2 + 1/3^2 + \dots \infty. \tag{4.15}$$

Solution. (i) The equations (4.3), (4.7) and (4.9) determine the Fourier coefficients:

$$a_0 = (1/2\pi) \int_{-\pi}^{\pi} (x + x^2)\, dx = (1/2\pi) \{ \int_{-\pi}^{\pi} x.\, dx + \int_{-\pi}^{\pi} x^2.\, dx \}$$

$$= (1/\pi) \{0 + \int_0^{\pi} x^2.\, dx\} = (1/\pi) \left[x^3/3\right]_0^{\pi} = \pi^2/3,$$

$$a_n = (1/\pi) \int_{-\pi}^{\pi} (x + x^2).\cos nx.dx$$

$$= (1/\pi) \{\int_{-\pi}^{\pi} x.\cos nx.\, dx + \int_{-\pi}^{\pi} x^2.\cos nx.\, dx\}$$

$$= (2/\pi) \{ 0 + \int_0^{\pi} x^2.\cos nx.dx \}$$

$$= (2/\pi) \{ \left[(x^2/n).\sin nx\right]_0^{\pi} - (2/n) \int_0^{\pi} x.\sin nx.\, dx \}$$

$$= (2\pi/n).\sin n\pi + (4/n\pi) \{\left[(x/n).\cos nx\right]_0^{\pi} - (1/n) \int_0^{\pi} \cos nx.\, dx\}$$

$$= (4/n^2).\cos n\pi - (4/n^3\pi) \left[\sin nx\right]_0^{\pi} = (4/n^2).\cos n\pi = (4/n^2).(-1)^n,$$

and

$$b_n = (1/\pi) \int_{-\pi}^{\pi} (x + x^2).\sin nx.dx$$

$$= (1/\pi) \{\int_{-\pi}^{\pi} x.\sin nx.\, dx + \int_{-\pi}^{\pi} x^2.\sin nx.\, dx\}$$

$$= (2/\pi) \{\int_0^{\pi} x.\sin nx.\, dx + 0\} = (2/n\pi)\{\left[-x.\cos nx\right]_0^{\pi} + \int_0^{\pi} \cos nx\, dx\}$$

$$= -(2/n).\cos n\pi + (2/n^2\pi)\left[\sin nx\right]_0^{\pi} = -(2/n).\cos n\pi = (2/n).(-1)^{n+1}.$$

Accordingly, the Fourier series in Eq. (3.2) reduces to

$$x + x^2 = \pi^2/3 + 4\{-(\cos x)/1^2 + (\cos 2x)/2^2 - (\cos 3x)/3^2 + \dots \infty\}$$

$$+ 2\{(\sin x)/1 - (\sin 2x)/2 + (\sin 3x)/3 - \dots \infty\}. \qquad (4.16)$$

(ii) Similarly, the Fourier series for the function x^2 in above interval is

$$x^2 = \pi^2/3 + 4\{-(\cos x)/1^2 + (\cos 2x)/2^2 - (\cos 3x)/3^2 + \dots \infty\}. \quad (4.17)$$

Putting $x = \pi$ in above result, we deduce

$$\pi^2 - \pi^2/3 = 2\pi^2/3 = 4\{1 + 1/2^2 + 1/3^2 + \dots \infty\},$$

which yields (4.15). //

Example 4.3. Obtain a Fourier series expansion of the function $f(x) = x.\sin x$ in the interval $[-\pi, \pi]$. Hence, deduce the result

$$\pi/4 = 1/2 + (1/1.3 - 1/3.5 + 1/5.7 - \dots \infty). \qquad (4.18)$$

Solution. (i) The function being even, the Fourier coefficients are evaluated by Eq. (4.12):

$$a_0 = (1/\pi)\{\int_0^\pi x.\sin x.\,dx\} = (1/\pi)\{[-x.\cos x]_0^\pi + \int_0^\pi \cos x.dx\}$$

$$= -\cos\pi + (1/\pi)[\sin x]_0^\pi = 1,$$

$$a_n = (2/\pi).\int_0^\pi x.\sin x.\cos nx.\,dx$$

$$= (1/\pi)\int_0^\pi x.\{\sin(n+1)x - \sin(n-1)x\}.\,dx$$

$$= (1/\pi)\{\int_0^\pi x.\sin(n+1)x.\,dx - \int_0^\pi x.\sin(n-1)x\}.\,dx \qquad (4.19)$$

$$= \{[-x.\cos(n+1)x]_0^\pi + \int_0^\pi \cos(n+1)x.\,dx\}/(n+1)\pi$$

$$+ \{[x.\cos(n-1)x]_0^\pi - \int_0^\pi \cos(n-1)x.\,dx\}/(n-1)\pi, \qquad \text{when } n \neq 1,$$

$$= [-\{\cos(n+1)\pi\}/(n+1) + \{\sin(n+1)\pi\}/(n+1)^2\pi]$$

$$+ \; [\{\cos (n-1)\,\pi\}/(n-1) - \{\sin (n-1)\,\pi\}/\,(n-1)^2\,\pi]$$

$$= (\cos n\pi)\;\{1/\,(n+1) - 1/\,(n-1)\} \;=\; -2\,(\cos n\pi)\,/\,(n-1)\,(n+1)$$

$$=\; 2\,(-1)^{n+1}/\,(n-1)\,(n+1),$$

and $b_n = 0$. However, when $n = 1$, Eq. (4.19) determines

$$a_1 \;=\; (1/\pi).\int_0^\pi \; x.\,\sin 2x.\,dx$$

$$=\; (1/2\pi)\;\{\;\left[-x.\cos 2x\right]_0^\pi + \int_0^\pi \;\cos 2x.\,dx\;\}$$

$$=\; -1/2\; + (1/4\pi)\;\left[\sin 2x\right]_0^\pi$$

$$=\; -1/2.$$

Accordingly, the Fourier series in Eq. (3.2) reduces to

$$x.\,\sin x \;=\; 1 - (\cos x)/2 + 2\sum_{n=2}^{\infty}\;\{(-1)^{n+1}\cos nx\}\,/\,(n-1)(n+1)$$

$$= 1 - (\cos x)/2 + 2\;\{(-\cos 2x)/1.3 + (\cos 3x)/2.4 - (\cos 4x)/3.5$$

$$+ (\cos 5x)\,/\,4.6 - (\cos 6x)\,/\,5.7 + \ldots\,\infty\}. \qquad (4.20)$$

(ii) Putting $x = \pi/2$ in above result, we deduce

$$\pi/2 \;=\; 1 + 2\;\{1/\,1.3 + 0 - 1/\,3.5 + 0 + 1/\,5.7 - 0 + \ldots\;\infty\},$$

that yields Eq. (4.18). //

Example 4.4. In the interval $[-\pi, \pi]$, prove that

$$x.\,(\pi^2 - x^2)\,/\,12 = \;(\sin x)/1^3 - (\sin 2x)/2^3 + (\sin 3x)/3^3 - \;\ldots\,\infty\}. \quad (4.21)$$

Solution. The function $x\,(\pi^2 - x^2)\,/\,12$ being odd the Fourier coefficients are evaluated by Eq (4.13):

$$a_0 \;=\; a_n \;=\; 0, \quad b_n \;=\; (2\,/\,12\pi)\int_0^\pi \; x.\,(\pi^2 - x^2).\,\sin nx.\,dx.$$

Carrying out the integration by the method of parts taking sin nx as the second function we obtain

$$b_n = (1/6n\pi).\ \{\left[-x.(\pi^2 - x^2).\cos nx\right]_0^\pi + \int_0^\pi (\pi^2 - 3x^2).\ \cos nx\ dx\}$$

$$= (1/6n^2\pi).\ \{\ \left[(\pi^2 - 3x^2).\sin nx\right]_0^\pi + 6\int_0^\pi x.\ \sin nx\ dx\}$$

$$= (1/n^3\pi).\ \{\ \left[-x.\cos nx\right]_0^\pi + \int_0^\pi \cos nx.\ dx\ \}$$

$$= (1/n^3)\ \{-\cos n\pi + (1/n\pi)\ \left[\sin nx\right]_0^\pi\} = -(1/n^3)\cos n\pi = (-1)^{n+1}/n^3.$$

Accordingly, the Fourier series in Eq. (3.2) for the function reads as

$$x.\ (\pi^2 - x^2)/12 = \sum_{n=1}^{\infty} \{(-1)^{n+1}/n^3\}.\ \sin nx$$

$$= (\sin x)/1^3 - (\sin 2x)/2^3 + (\sin 3x)/3^3 - \ldots \infty\}.\ //$$

§ 5. Fourier Series in Any Interval

5.1. Interval [– l, l]: The equation (4.11) gives a Fourier series expansion for the function $f(x)$ in the interval $(-\pi, \pi)$. Analogously, for some function $F(y)$ also defined in the same interval $[-\pi, \pi]$, Eq. (4.11) reads as

$$F(y) = (1/2\pi)\int_{-\pi}^{\pi} F(t)\,dt + (1/\pi)\sum_{n=1}^{\infty}\int_{-\pi}^{\pi} F(t)\cos\{n(t-y)\}dt. \quad (5.1)$$

Applying the change of parameters:

$$t = \pi u/l \quad \Rightarrow \quad dt = (\pi.\ du)/l, \quad \text{and} \quad y = \pi x/l. \quad (5.2)$$

Eq. (5.1) assumes the form

$$F(\pi x/l) = (1/2l)\int_{-l}^{l} F(\pi u/l)\,du$$

$$+ (1/l)\sum_{n=1}^{\infty}\int_{-l}^{l} F(\pi u/l)\cos\{n\pi(u-x)/l\}.\ du$$

Writing $F(\pi x/l)$ as $f(x)$, and similarly $F(\pi u/l)$ as $f(u)$ above relation further alters as

$$f(x) = (1/2l)\int_{-l}^{l} f(u)\,du + (1/l)\sum_{n=1}^{\infty}\int_{-l}^{l} f(u)\cos\{n\pi(u-x)/l\}.du$$

$$= (1/2l)\int_{-l}^{l} f(u)\,du + (1/l)\sum_{n=1}^{\infty}\int_{-l}^{l} f(u)\,\{\cos(n\pi u/l)\cos(n\pi x/l)$$

$$+ \sin(n\pi u/l)\sin(n\pi x/l)\}\,du, \tag{5.3}$$

which is the desired Fourier series expansion of $f(x)$ in interval $[-l, l]$ together with the Fourier coefficients:

$$\left.\begin{aligned} a_0 &= (1/2l)\int_{-l}^{l} f(u)\,du,\quad a_n = (1/l)\int_{-l}^{l} f(u)\cos(n\pi u/l)\,du \\ b_n &= (1/l)\int_{-l}^{l} f(u)\sin(n\pi u/l)\,du. \end{aligned}\right\} \tag{5.4}$$

5.2. Interval [0, π]: Let a function $f(x)$ be defined on the interval $[0, \pi]$. In order to obtain a *cosine* series expansion for it on this interval, we construct a new function, say $g(x)$ on the extended interval $[-\pi, \pi]$ as follows:

$$g(x) = f(-x) \text{ (when } -\pi \le x < 0\text{), and } f(x) \text{ (when } 0 \le x \le \pi\text{)}. \tag{5.5}$$

As such, the new function is even on the extended interval and agrees with the given function $f(x)$ on the interval $[0, \pi]$. Thus, $g(x)$ can be expanded in a Fourier series as discussed in Sub-section 4.4, Case (i). The Fourier coefficients for the same are

$$a_0 = (1/2\pi)\int_{-\pi}^{\pi} g(x)\,dx = (1/\pi)\int_{0}^{\pi} g(x)\,dx = (1/\pi)\int_{0}^{\pi} f(x)\,dx$$

$$a_n = (1/\pi)\int_{-\pi}^{\pi} g(x)\cos nx.\,dx = (2/\pi)\int_{0}^{\pi} g(x).\cos nx.\,dx$$

$$= (2/\pi)\int_{0}^{\pi} f(x).\cos nx\,dx,$$

and

$$b_n = (1/\pi)\int_{-\pi}^{\pi} g(x).\sin nx\,dx = 0,$$

as the integrand $g(x).\sin nx$ then becomes an odd function of x causing vanishing of the last integral. We note that above values of the Fourier coefficients are in accordance with Eq. (4.12). Consequently, the Fourier series for $g(x)$ on the interval $[-\pi, \pi]$ turns out to be a purely *cosine* series given by Eq. (4.10a) on $[-\pi, \pi]$. For $g(x)$ being same as $f(x)$ on the interval $[0, \pi]$ the same *cosine* series is valid for $f(x)$ too on the interval $[0, \pi]$.

Example 5.1. Find a series of cosines of multiples of x which may represent the function $f(x) = x$ on the interval $[0, \pi]$.

Solution. To have a cosine series only the function must be so defined that it be *even* in the extended interval $[-\pi, \pi]$. Therefore, as seen above the Fourier coefficients are determined as per Eq. (4.12):

$$a_0 = (1/\pi)\int_0^\pi f(x)\,dx = (1/\pi)\int_0^\pi x\,dx = (1/\pi)\left[x^2/2\right]_0^\pi = \pi/2,$$

$$a_n = (2/\pi)\int_0^\pi x\cos nx.\,dx = (2/n\pi)\{\left[x.\sin nx\right]_0^\pi - \int_0^\pi \sin nx.\,dx\}$$

$$= (2/n^2\pi)\left[\cos nx\right]_0^\pi = (2/n^2\pi)(\cos n\pi - 1) = (2/n^2\pi)\{(-1)^n - 1\}$$

$$= 0 \text{ (when } n \text{ is even)}, \quad \text{and } -4/n^2\pi \quad \text{(when } n \text{ is odd)},$$

and $b_n = 0$. Thus, the Fourier series reducing to cosine series, given by Eq. (4.10a), reads as

$$x = \pi/2 - (4/\pi).\{(\cos x)/1^2 + (\cos 3x)/3^2 + (\cos 5x)/5^2 + \ldots \infty\}. \quad (5.6)$$

Particularly, when $x = 0$, above series implies the result

$$\pi^2/8 = 1/1^2 + 1/3^2 + 1/5^2 + \ldots \infty. //$$

Example 5.2. Obtain a cosine series for the function $f(x) = \sin x$ on the interval $[0, \pi]$.

Solution. In order to have a cosine series expansion the function must be so defined that it be *even* in the extended interval $[-\pi, \pi]$. Therefore, as seen above the Fourier coefficients are determined as per Eq. (4.12):

$$a_0 = (1/\pi)\int_0^\pi \sin x.\,dx = (1/\pi)\left[-\cos x\right]_0^\pi = (1/\pi)(-\cos\pi + 1) = 2/\pi,$$

$$a_n = (2/\pi) \int_0^\pi \sin x . \cos nx . dx$$

$$= (1/\pi) \int_0^\pi \{\sin (n+1) x - \sin (n-1) x\} dx \qquad (5.7)$$

$$= (1/\pi) \big[\{-\cos (n+1) x\}/(n+1) + \{\cos (n-1) x\}/(n-1)\big]_0^\pi, \text{ when } n \neq 1$$

$$= (1/\pi) \big[-\{\cos (n+1) \pi\}/(n+1) + \{\cos (n-1) \pi\}/(n-1) + \{1/(n+1) - 1/(n-1)\}\big]$$

$$= (1/\pi) \big[(\cos n\pi)/(n+1) - (\cos n\pi)/(n-1) + \{1/(n+1) - 1/(n-1)\}\big]$$

$$= (1/\pi) [\{(-1)^n + 1\}\{1/(n+1) - 1/(n-1)\}] = -2\{(-1)^n + 1\}/(n-1)(n+1)\pi$$

$= 0$ (when n is odd), and $-4/(n-1)(n+1)\pi$ (when n is even),

and $b_n = 0$.

On the other hand, when $n = 1$, Eq. (5.7) determines

$$a_1 = (1/\pi) \int_0^\pi \sin 2x . dx = (1/2\pi) \big[-\cos 2x\big]_0^\pi = (-1+1)/2\pi = 0.$$

Accordingly, the *cosine* series given by Eq. (4.10a) assumes the form

$$\sin x = 2/\pi - (2/\pi) \sum_{n=2}^{\infty} [\{(-1)^n + 1\}/(n-1)(n+1)] \cos nx, \; n \text{ being even}$$

$$= 2/\pi - (4/\pi) \sum_{m=1}^{\infty} (\cos 2mx)/(2m-1)(2m+1), \text{ where } n = 2m$$

$$= 2/\pi - (4/\pi) \{(\cos 2x)/1.3 + (\cos 4x)/3.5 + (\cos 6x)/5.7$$

$$+ (\cos 8x)/7.9 + \ldots \infty\}. //$$

Example 5.3. Obtain the Fourier series for the function $f(x) = \cos x$ for all values of x in the interval $[0, \pi]$.

Solution. We construct a new function $g(x)$ in the extended interval $[-\pi, \pi]$ such that

$$g(x) = -\cos x \;\; (\text{when } -\pi \leq x \leq 0), \;\; \cos(x) \;\; (\text{when } 0 \leq x \leq \pi).$$

As per definition it is an odd function of x for which the Fourier coefficients are given by Eq. (4.13): $a_0 = a_n = 0$, and

$$b_n = (2/\pi) \int_0^{\pi} \cos x.\ \sin nx.\ dx$$

$$= (1/\pi) \int_0^{\pi} \{\sin (n+1)\, x + \sin (n-1)\, x\}.dx \qquad (5.8)$$

$$= (1/\pi)\left[-\{\cos(n+1)\,x\}/(n+1) - \{\cos\ (n-1)\,x\}/\,(n-1)\right]_0^{\pi},\ \text{when } n \neq 1$$

$$= \left[-\{\cos(n+1)\,\pi\}/(n+1) - \{\cos\ (n-1)\,\pi\}/\,(n-1) + \{1/(n+1) + 1/(n-1)\}\right]/\pi$$

$$= (1/\pi)\left[(\cos n\pi)/\,(n+1)\ + (\cos\ n\pi)\,/\,(n-1) + 1/(n+1) + 1/(n\text{-}1)\right]$$

$$= (1/\pi)\ [\ \{(-1)^n + 1\}\{1\,/\,(n+1) + 1\,/\,(n-1)\}\ ]$$

$$= (2n\,/\,\pi)\ \{\ (-1)^n + 1\ \}\,/\,(n-1)\,(n+1)$$

$$= 0 \text{ (when } n \text{ is odd), and } 4\,/\,(n-1)(n+1)\,\pi \text{ (when } n \text{ is even).} \qquad (5.9)$$

On the other hand, when $n = 1$, Eq. (5.8) determines

$$b_1 = (1/\pi)\int_0^{\pi} \sin 2x.dx = (1/2\pi)\left[-\cos 2x\right]_0^{\pi} = (-1+1)\,/\,2\pi = 0.$$

Accordingly, the cosine series given by Eq. (4.10b) assumes the form

$$\cos x = \sum_{n=2}^{\infty} b_n.\sin nx = (4/\pi)\sum_{n=2}^{\infty} (n \sin nx)/(n-1)(n+1), \text{ by Eq. (5.9)}$$

(where only even values of n give non-zero terms)

$$= (4/\pi)\{2\ (\sin 2x)\,/1.3 + 4\ (\sin 4x)/3.5 + 6\ (\sin 6x)/5.7 + \ldots \infty\}.\ //$$

Note 5.1. Defining $g\ (x)$ as even function on the interval $[-\pi, \pi]$ in analogy with Eq. (5.5), we note that all the Fourier coefficients vanish for $\cos x$ except a_1:

$$a_0 = (1/\pi) \int_0^{\pi} \cos x.\ dx = (1/\pi)\ \left[\sin x\right]_0^{\pi} = 0,$$

$$a_n = (2/\pi) \int_0^{\pi} \cos x.\ \cos nx.\ dx$$

$$= (1/\pi) \int_0^{\pi} \{\cos (n+1)\, x + \cos (n-1)\, x\}\, dx$$

$$= (1/\pi)\big[\{\sin (n+1)\, x\}/(n+1) + \{\sin (n-1)\, x\}/(n-1)\big]_0^{\pi} = 0, \text{ when } n \neq 1;$$

$$a_1 = (1/\pi)\int_0^{\pi} (\cos 2x + 1)\, dx = (1/\pi)\big[(\sin 2x)/2 + x\big]_0^{\pi} = 1$$

and $b_n = 0$. Thus, a *cosine* series for $\cos x$ reduces to just $\cos x$.

5.3. A sine series in the interval [0, π]: To obtain a *sine* series expansion for a function $f(x)$ defined on the interval $[0, \pi]$, we must construct a new function, say $g(x)$ on the extended interval $[-\pi, \pi]$ which should be odd:

$$g(x) = f(x) \text{ (when } 0 \leq x \leq \pi) \text{ and } -f(x) \text{ (when } -\pi \leq x < 0), \quad (5.10)$$

and agree with the given function $f(x)$ on the interval $[0, \pi]$. The Fourier coefficients for the same are given by Eq. (4.13). Accordingly, its Fourier series expansion turns out to be purely a *sine* series as in Eq. (4.10b). Since $g(x)$, as per definition, agrees with $f(x)$ on the interval $[0, \pi]$ its Fourier series in the interval $[0, \pi]$ becomes the same as for $f(x)$.

Example 5.4. Expand the function $f(x) = \sin x$ in a sine series on the interval $[0, \pi]$.

Solution. We construct a new function $g(x)$ on the extended interval $[-\pi, \pi]$ as in Eq. (5.10). The function being odd in $[-\pi, \pi]$ its Fourier series becomes purely a *sine* series given by Eq. (4.10b):

$$g(x) = \sum_{n=1}^{\infty} b_n .\sin nx \qquad (5.11)$$

and the coefficients b_n are given by Eq. (4.13):

$$b_n = (2/\pi)\int_0^{\pi} g(x).\sin nx.\, dx.$$

For $g(x) = f(x)$ on the interval $[0, \pi]$, above value becomes

$$b_n = (2/\pi)\int_0^{\pi} f(x).\sin nx.\, dx = (2/\pi)\int_0^{\pi} \sin x.\sin nx.\, dx$$

$$= (1/\pi)\int_0^\pi \{\cos (n-1)x - \cos (n+1)x\}.dx \qquad (5.12)$$

$$= (1/\pi)\left[\{\sin (n-1)x\}/(n-1) - \{\sin (n+1)x\}/(n+1)\right]_0^\pi = 0, \text{ when } n \neq 1.$$

On the other hand, when $n = 1$, (5.12) determines

$$b_1 = (1/\pi)\int_0^\pi (1 - \cos 2x).dx = (1/\pi)\left[x - (\sin 2x)/2\right]_0^\pi = 1.$$

Thus, the RHS of Eq. (5.11) reduces to just one term, sin x, which is the value of the function itself. //

5.4. The interval [0, l]: As seen in the Sub-section 5.2, a function can be expanded in the interval [0, π] in a *cosine* series of the form in Eq. (4.10a). Re-writing it for a function $F(y)$:

$$F(y) = (1/\pi)\,[\int_0^\pi F(t)\,dt + 2\sum_{n=1}^{\infty} \{\int_0^\pi F(t)\cos nt.\,dt\}\cos ny]. \quad (5.13)$$

Applying the change of parameters as in Eq. (5.2), above result assumes the form:

$$F(\pi x/l) = (1/l)\,[\int_0^l F(\pi u/l)\,du$$

$$+ 2\sum_{n=1}^{\infty} \{\int_0^l F(\pi u/l).\cos (n\pi u/l)\,du\}.\cos (n\pi x/l)\,]$$

Writing $F(\pi x/l) \equiv f(x)$, and $F(\pi u/l) \equiv f(u)$, above relation further alters as

$$f(x) = (1/l)\,[\int_0^l f(u)\,du$$

$$+ 2\sum_{n=1}^{\infty} \{\int_0^l f(u)\cos (n\pi u/l)\,du\}\cos (n\pi x/l)\,], \qquad (5.14)$$

which is a *cosine* series.

The *sine* series expansion of a function $f(x)$ in the interval [0, π] is discussed in the Sub-section 5.3 and it is given by Eq. (4.10b). Re-writing it for a function $F(y)$:

$$F(y) = (2/\pi)\sum_{n=1}^{\infty} \{\int_0^\pi F(t)\sin nt.\,dt\}.\sin ny, \qquad (5.15)$$

and applying the change of parameters given by Eq. (5.2), above equation reduces to

$$F(\pi x/l) = (2/l)\sum_{n=1}^{\infty} \{\int_0^l F(\pi u/l)\sin(n\pi u/l)\,du\}\sin(n\pi x/l)$$

or, $$f(x) = (2/l)\sum_{n=1}^{\infty} \{\int_0^l f(u)\sin(n\pi u/l)\,du\}\sin(n\pi x/l). \;// \quad (5.16)$$

Example 5.5. For all values of x in the interval $[-\pi/2, \pi/2]$, prove

$$x = (4/\pi)\{(\sin x)/1^2 - (\sin 3x)/3^2 + (\sin 5x)/5^2 - \ldots \infty\}. \quad (5.17)$$

Solution. Setting

$$x + \pi/2 = z \quad (5.18)$$

so that when x varies from $-\pi/2$ to $\pi/2$ the new variable z varies from 0 to π. The *cosine* series expansion of x in the interval $[0, \pi]$ is obtained vide equation (5.6). Analogously, we can write a *cosine* series for z in the interval $[0, \pi]$:

$$z = \pi/2 - (4/\pi)\{(\cos z)/1^2 + (\cos 3z)/3^2 + (\cos 5z)/5^2 + \ldots \infty\},$$

which, for Eq. (5.18), takes the form as given by Eq. (5.17). //

5.5. The interval [0, 2π]: Putting $t = u - \pi$ so that $dt = du$ and $y = x - \pi$ in Eq. (5.1) we deduce

$$F(x-\pi) = (1/2\pi)\int_{u=0}^{2\pi} F(u-\pi)\,du$$

$$+ (1/\pi)\sum_{n=1}^{\infty}\int_{u=0}^{2\pi} F(u-\pi)\cos\{n(u-x)\}du.$$

Writing $F(x-\pi)$ as $f(x)$, and $F(u-\pi)$ as $f(u)$ above relation further alters as

$$f(x) = (1/2\pi)\int_{u=0}^{2\pi} f(u)\,du + (1/\pi)\sum_{n=1}^{\infty}\int_{u=0}^{2\pi} f(u)\cos\{n(u-x)\}\,du \quad (5.19a)$$

$$= (1/2\pi)\int_{u=0}^{2\pi} f(u)\,du$$

$$+ (1/\pi)\sum_{n=1}^{\infty}\int_{u=0}^{2\pi} f(u)\{\cos nu.\cos nx + \sin nu.\sin nx)\}\,du$$

$$= (1/2\pi)\int_{u=0}^{2\pi} f(u)\,du + (1/\pi)\sum_{n=1}^{\infty} [\{\int_{u=0}^{2\pi} f(u).\cos nu.\,du\}\cos nx$$

$$+ \{\int_{u=0}^{2\pi} f(u).\,\sin nu\,.du\}\,\sin nx]. \qquad (5.19b)$$

Note 5.2. Comparing the results given by Eqs. (5.19a) and (5.19b) with Eqs. (4.11) and (4.10) respectively, we note that the difference lies in the limits only.

5.6. The interval [0, 1]: Rewriting the Fourier series expansion given by Eq. (5.19a) for a function $F(y)$ defined in the interval $[0, 2\pi]$:

$$F(y) = (1/2\pi)\int_{t=0}^{2\pi} F(t)\,dt$$

$$+ (1/\pi)\sum_{n=1}^{\infty} \int_{t=0}^{2\pi} F(t)\cos\{n(t-y)\}dt, \qquad (5.20)$$

and applying the change of parameters $t = 2\pi u$ so that $dt = 2\pi\,du$ and $y = 2\pi x$, we derive

$$F(2\pi x) = \int_{u=0}^{1} F(2\pi u)\,du + 2\sum_{n=1}^{\infty} \int_{u=0}^{1} F(2\pi u)\cos\{2n\pi(u-x)\}du.$$

Setting $F(2\pi x) = f(x)$ and $F(2\pi u) = f(u)$, above expansion reads as

$$f(x) = \int_{u=0}^{1} f(u)\,du + 2\sum_{n=1}^{\infty} \int_{u=0}^{1} f(u)\cos\{2n\pi(u-x)\}\,du$$

$$= \int_{u=0}^{1} f(u)\,du + 2\sum_{n=1}^{\infty} [\{\int_{u=0}^{1} f(u)\cos(2n\pi u).\,du\}\cos(2n\pi x)$$

$$+ \{\int_{u=0}^{1} f(u)\sin(2n\pi u).du\}.\sin(2n\pi x)]. \qquad (5.21)$$

Particularly, the *cosine* (respectively *sine*) series expansions for a function defined in the interval [0, 1] can be found in analogy with Eq. (5.14) respectively Eq. (5.16):

$$f(x) = \int_{u=0}^{1} f(u)\,du$$

$$+ 2\sum_{n=1}^{\infty} \{\int_{u=0}^{1} f(u)\cos(n\pi u)\,du\}.\,\cos(n\pi x) \qquad (5.22)$$

(respectively)

$$f(x) = 2\sum_{n=1}^{\infty} \{\int_{u=0}^{1} f(u)\sin(n\pi u)\,du\}.\ \sin(n\pi x).// \qquad (5.23)$$

Example 5.6. Find the Fourier series expansion for the function $f(x) = e^{2x}$ defined on the interval [0, 1].

Solution. The Fourier coefficients in the expansion given by Eq. (5.21) are:

$$a_0 = \int_{x=0}^{1} e^{2x}\,dx = \left[e^{2x}/2\right]_0^1 = (e^2 - 1)/2,$$

$$a_n = 2\int_{x=0}^{1} e^{2x}.\cos(2n\pi x)\,dx = \left[e^{2x}(\cos 2n\pi x + n\pi.\sin 2n\pi x)\right]_0^1/(n^2\pi^2 + 1)$$

$$= (e^2.\cos 2n\pi - 1)/(n^2\pi^2 + 1) = (e^2 - 1)/(n^2\pi^2 + 1),$$

and

$$b_n = 2\int_{x=0}^{1} e^{2x}.\sin(2n\pi x)\,dx = \left[e^{2x}(\sin 2n\pi x - n\pi.\cos 2n\pi x)\right]_0^1/(n^2\pi^2 + 1)$$

$$= n\pi(-e^2.\cos 2n\pi + 1)/(n^2\pi^2 + 1) = -n\pi(e^2 - 1)/(n^2\pi^2 + 1),$$

where we have used the integral formulae given by Eqs. (2.5.17) and (2.5.19). Thus, the desired Fourier series is

$$e^{2x} = (e^2 - 1)\,[1/2 + \sum_{n=1}^{\infty} \{\cos(2n\pi x) - n\pi.\sin(2n\pi x)\}/(n^2\pi^2 + 1)].(5.24)$$

5.7. The interval [*a*, *b*]: Applying the change of parameters

$$t = 2\pi(u - a)/(b - a) \Rightarrow dt = 2\pi\,du/(b - a) \text{ and } y = 2\pi(x - a)/(b - a),$$

the equation (5.20) reads as

$$F\{2\pi(x - a)/(b - a)\} = [\int_{u=a}^{b} F\{2\pi(u - a)/(b - a)\}\,du$$

$$+ 2\sum_{n=1}^{\infty} \int_{u=a}^{b} F\{2\pi(u - a)/(b - a)\}\cos\{2n\pi(u - x)/(b - a)\}du]/(b - a)$$

or,
$$f(x) = [\int_{u=a}^{b} f(u)\,du$$

$$+ 2\sum_{n=1}^{\infty} \int_{u=a}^{b} f(u) \cos \{2n\pi (u - x) / (b - a)\} du \,] / (b - a). \qquad (5.25)$$

§ 6. Fourier Series for Piecewise Defined Functions

Let a function $f(x)$ be defined on some closed interval $[a, b]$ at points of the interval except for a finite number of points, say $x_1, x_2, \ldots, x_n$. It is called *piecewise continuous* on the interval $[a, b]$ if:

(i) it is continuous on each sub-intervals $(a, x_1), (x_1, x_2), \ldots, (x_{n-1}, x_n), (x_n, b)$;

(ii) it possesses a finite limit from the right at the (left) end $x = a$, a finite limit from the left at the (right) end $x = b$, and both left and right limits at each point x_i, $i = 1, 2, \ldots, n$.

We shall obtain Fourier series expansions for such functions in the following. The method is demonstrated by means of some examples.

Example 6.1. Find the Fourier series for the function defined on the interval $[-2, 2]$:

$$f(x) = 0 \text{ (when } -2 \le x < 0), \; 1 \text{ (when } 0 \le x < 1), \; 2 \text{ (when } 1 \le x \le 2).$$

Solution. A Fourier series expansion of a function defined over the interval $[-l, l]$ is obtained by Eq. (5.3) and the corresponding Fourier coefficients by Eq. (5.4). Thus, for $l = 2$, we have

$$a_0 = (1/4)\int_{-2}^{2} f(x)\, dx = (1/4)\,\{\int_{-2}^{0} + \int_{0}^{1} + \int_{1}^{2}\}\, f(x)\, dx$$

$$= (1/4)\,\{\, 0 + \int_{0}^{1} dx + 2\int_{1}^{2} dx\} = (1/4)\,\{\, [x]_0^1 + 2\, [x]_1^2 \,\}$$

$$= (1/4)\,(1 + 2) = 3/4,$$

$$a_n = (1/2)\int_{-2}^{2} f(x) \cos (n\pi x/2)\, dx$$

$$= (1/2)\,\{\int_{-2}^{0} + \int_{0}^{1} + \int_{1}^{2}\}.\, f(x) \cos (n\pi x/2)\, dx$$

$$= (1/2)\ \{\int_0^1 \cos(n\pi x/2)\,dx + 2\int_1^2 \cos(n\pi x/2)\,dx\}$$

$$= (1/n\pi)\ \{\ [\sin(n\pi x/2)]_0^1 + 2\ [\sin(n\pi x/2)]_1^2\ \}$$

$$= (1/n\pi)\ [\sin(n\pi/2) + 2\{\sin n\pi - \sin(n\pi/2)\}] = -\ \{\sin(n\pi/2)\}/n\pi,$$

and

$$b_n = (1/2)\int_{-2}^2 f(x)\sin(n\pi x/2)\,dx$$

$$= (1/2)\ \{\int_{-2}^0 + \int_0^1 + \int_1^2\ \}.\,f(x)\sin(n\pi x/2)\,dx$$

$$= (1/2)\ \{\int_0^1 \sin(n\pi x/2)\,dx + 2\int_1^2 \sin(n\pi x/2)\,dx\ \}$$

$$= -\ (1/n\pi)\ \{\ [\cos(n\pi x/2)]_0^1 + 2\ [\cos(n\pi x/2)]_1^2\ \}$$

$$= -(1/n\pi)\ [\ \cos(n\pi/2) - 1 + 2\ \{\ \cos(n\pi) - \cos(n\pi/2)\ \}\ ]$$

$$= (1/n\pi)\ \{\ 1 + \cos(n\pi/2) - 2.\ \cos n\pi\ \}.$$

Thus, the desired Fourier series for above function is

$$f(x) = 3/4 + (1/n\pi).\sum_{n=1}^{\infty}\ [-\sin(n\pi/2).\ \cos(n\pi x/2)$$

$$+ \{\ 1 + \cos(n\pi/2) - 2.\cos n\pi\ \}.\ \sin(n\pi x/2)\ ].\ //$$

Example 6.2. Find the Fourier series for the function defined by

$$f(x) = -1\ (\text{when } -1 \le x < 0),\ \ 1\ (\text{ when } 0 \le x \le 1)$$

on the interval [– 1, 1].

Solution. The equation (5.4), for $l = 1$, determines the Fourier coefficients:

$$a_0 = (1/2)\int_{-1}^1 f(x)\,dx = (1/2)\ \{\int_{-1}^0 + \int_0^1\ \}f(x)\,dx$$

$$= (1/2)\ \{-\int_{-1}^0 dx + \int_0^1 dx\} = (1/2)\{[x]_0^{-1} + [x]_0^1\ \} = (1/2)\,(-1+1) = 0,$$

$$a_n = \int_{-1}^{1} f(x) \cos (n\pi x)\, dx = \{\int_{-1}^{0} + \int_{0}^{1} \} f(x) \cos (n\pi x)\, dx$$

$$= -\int_{-1}^{0} \cos (n\pi x)\, dx + \int_{0}^{1} \cos (n\pi x)\, dx$$

$$= (1/n\pi) \{ [\sin (n\pi\, x)]_0^{-1} + [\sin (n\pi\, x)]_0^1 \} = 0,$$

and

$$b_n = \int_{-1}^{1} f(x).\, \sin (n\pi x)\, dx = \{\int_{-1}^{0} + \int_{0}^{1} \} f(x).\, \sin (n\pi x)\, dx$$

$$= -\int_{-1}^{0} \sin (n\pi x)\, dx + \int_{0}^{1} \sin (n\pi x)\, dx$$

$$= \{[\cos (n\pi\, x)]_{-1}^{0} - [\cos (n\pi\, x)]_0^1 \}/n\pi = (1 - \cos n\pi - \cos n\pi + 1) / n\pi$$

$$= 2\,(1 - \cos n\pi) / n\pi = 0 \qquad \text{(for even } n), \qquad 4 / n\pi \quad \text{(for odd } n).$$

Thus, the desired Fourier series for above function is

$$f(x) = 2 \sum_{n=1}^{\infty} \{ (1 - \cos n\pi) / n\pi \} \sin (n\pi x)$$

$$= (4/\pi) \sum_{m=1}^{\infty} \sin \{ (2m - 1)\, \pi x) \} / (2m - 1), \quad n \text{ being odd, say } 2m - 1.\ //$$

Example 6.3. Find the Fourier series expansion for the function defined by

$$f(x) = k\,x \ \text{(when } 0 \le x \le l/2),\ k\,(l - x) \text{ (when } l/2 \le x \le l)$$

on the interval $[0, l\,]$.

Solution. (i) The *cosine* series expansion of a function defined on the interval $[0, l\,]$ is given by the equation (5.14). Breaking the range of integration therein, we have

$$a_0 = (1/l) \int_0^l f(x)\, dx = (1/l) \{\int_0^{l/2} + \int_{l/2}^{l} \} f(x)\, dx$$

$$= (k/l) \{ \int_0^{l/2} x\, dx + \int_{l/2}^{l} (l - x)\, dx \}$$

$$= (k/l) \{ [x^2/2]_0^{l/2} + [lx - x^2/2]_{l/2}^{l} \}$$

$$= (k/l)(l^2/8 + l^2 - l^2/2 - l^2/2 + l^2/8) = kl/4,$$

and

$$a_n = (2/l)\int_0^l f(x)\cos(n\pi x/l)\,dx$$

$$= (2/l)\{\int_0^{l/2} + \int_{l/2}^{l}\} f(x)\cos(n\pi x/l)\,dx$$

$$= (2k/l)\{\int_0^{l/2} x.\cos(n\pi x/l)\,dx + \int_{l/2}^{l}(l-x)\cos(n\pi x/l)\,dx\}$$

$$= (2k/n\pi)\{[x\sin(n\pi x/l)]_0^{l/2} - \int_0^{l/2}\sin(n\pi x/l)\,dx$$

$$+ [(l-x)\sin(n\pi x/l)]_{l/2}^{l} + \int_{l/2}^{l}\sin(n\pi x/l)\,dx\}$$

$$= (2k/n\pi)\{(l/2)\sin(n\pi/2) + (l/n\pi)[\cos(n\pi x/l)]_0^{l/2}$$

$$-(l/2)\sin(n\pi/2) - (l/n\pi)[\cos(n\pi x/l)]_{l/2}^{l}\}$$

$$= (2kl/n^2\pi^2)\{\cos(n\pi/2) - 1 - \cos n\pi + \cos(n\pi/2)\}$$

$$= (2kl/n^2\pi^2)\{2\cos(n\pi/2) - 1 - \cos n\pi\}$$

$$= (2kl/n^2\pi^2)\{2\cos(n\pi/2)(1 - \cos(n\pi/2)\},$$

which vanishes for all odd values of n as well as for even values of $n/2$, i.e. when $n = 4, 8, 12, 16, \ldots$ But, for odd values of $n/2$, i.e. when $n = 2, 6, 10, 14, \ldots$ the expression within the curly brackets is -4 determining the following values of a_n :

$$a_n = -8kl/n^2\pi^2, \text{ for } n = 2, 6, 10, 14, \ldots$$

Accordingly, the desired Fourier series is

$$f(x) = kl/4 - (8kl/\pi^2)\ [\{\cos(2\pi x/l)\}/2^2 + \{\cos(6\pi x/l)\}/6^2$$

$$+ \{\cos(10\pi x/l)\}/10^2 + \ldots].$$

(ii) On the other hand, the *sine* series expansion of a function defined on the interval $[0, l]$ is given by the equation (5.16). Proceeding similarly, we evaluate

$$\int_0^l f(x).\sin(n\pi x/l)\,dx = \{\int_0^{l/2} + \int_{l/2}^{l}\} f(x)\sin(n\pi x/l)\,dx$$

$$= k\{\int_0^{l/2} x.\sin(n\pi x/l)\,dx + \int_{l/2}^{l} (l-x)\sin(n\pi x/l)\,dx\}$$

$$= (kl/n\pi)\{\left[-x\cos(n\pi x/l)\right]_0^{l/2} + \int_0^{l/2} \cos(n\pi x/l)\,dx$$

$$-\left[(l-x)\cos(n\pi x/l)\right]_{l/2}^{l} - \int_{l/2}^{l} \cos(n\pi x/l)\,dx\}$$

$$= (kl/n\pi)\{-(l/2)\cos(n\pi/2) + (l/n\pi)\left[\sin(n\pi x/l)\right]_0^{l/2}$$

$$+ (l/2)\cos(n\pi/2) - (l/n\pi)\left[\sin(n\pi x/l)\right]_{l/2}^{l}\}$$

$$= (kl^2/n^2\pi^2)\{\sin(n\pi/2) - \sin n\pi + \sin(n\pi/2)\} = (2kl^2/n^2\pi^2)\sin(n\pi/2),$$

which vanishes for even values of n. Accordingly, the expansion (5.16) reduces to

$$f(x) = (4kl/\pi^2)\sum_{n=1}^{\infty} \{\sin(n\pi/2).\sin(n\pi x/l)\}/n^2$$

$$= (4kl/\pi^2)\,[\{\sin(\pi x/l)\}/1^2 - \{\sin(3\pi x/l)\}/3^2 + \{\sin(5\pi x/l)\}/5^2 - \ldots\}. //$$

Example 6.4. Find the Fourier series expansion for the following function defined on the interval $[-2l, 2l]$:

$$f(x) = \begin{cases} l & (\text{when } -2l \le x \le -l), \\ -x & (\text{when } -l \le x \le 0), \\ x & (\text{when } 0 \le x \le l), \\ l & (\text{when } l \le x \le 2l). \end{cases} \qquad (6.1)$$

Solution. The equation (5.1) gives the Fourier series expansion of a function $F(y)$ on the interval $[-\pi, \pi]$. Applying the change of parameters as:

$$t = \pi u/2l \;\Rightarrow\; dt = (\pi\,du)/2l, \text{ and } \quad y = \pi x/2l.$$

Eq. (5.1) assumes the form

$$F(\pi x / 2l) = (1/4l) \int_{-2l}^{2l} F(\pi u / 2l)\, du$$

$$+ (1/2l) \sum_{n=1}^{\infty} \int_{-2l}^{2l} F(\pi u / 2l) \cos\{n\pi (u-x)/2l\}\, du$$

Writing $F(\pi x / 2l)$ as $f(x)$, and similarly $F(\pi u/2l)$ as $f(u)$ above relation further alters as

$$f(x) = (1/4l) \int_{-2l}^{2l} f(u)\, du$$

$$+ (1/2l) \sum_{n=1}^{\infty} \int_{-2l}^{2l} f(u) \cos\{n\pi (u-x)/2l\}\, du. \qquad (6.2)$$

Since

$$\int_{-2l}^{2l} f(u)\, du = \{\int_{-2l}^{-l} + \int_{-l}^{0} + \int_{0}^{l} + \int_{l}^{2l}\} f(u)\, du$$

$$= \{ l \int_{-2l}^{-l} - \int_{-l}^{0} u + \int_{0}^{l} u + l \int_{l}^{2l} \}\, du$$

$$= l(-l+2l) + l^2/2 + l^2/2 + l(2l-l) = 3l^2,$$

and

$$\int_{-2l}^{2l} f(u) \cos\{ n\pi (u-x)/2l \}\, du$$

$$= \{ \int_{-2l}^{-l} + \int_{-l}^{0} + \int_{0}^{l} + \int_{l}^{2l} \} f(u) \cos\{n\pi (u-x)/2l\}\, du$$

$$= l \int_{-2l}^{-l} \cos\{n\pi (u-x)/2l\}\, du - \int_{-l}^{0} u.\cos\{n\pi (u-x)/2l\}\, du$$

$$+ \int_{0}^{l} u.\cos\{ n\pi (u-x)/2l \}\, du + l \int_{l}^{2l} \cos\{ n\pi (u-x)/2l \}\, du$$

$$= (2l/n\pi) \{ l \left[\sin\{n\pi(u-x)/2l\}\right]_{-2l}^{-l} + \left[u \sin\{n\pi(u-x)/2l\}\right]_{0}^{-l}$$

$$- \int_{0}^{-l} \sin\{n\pi (u-x)/2l\}\, du + \left[u \sin\{n\pi(u-x)/2l\}\right]_{0}^{l}$$

$$- \int_{0}^{l} \sin\{n\pi (u-x)/2l\}\, du + l \left[\sin\{n\pi(u-x)/2l\}\right]_{l}^{2l} \}$$

$$= (2l^2/n\pi)(-\sin\{ n\pi (l+x)/2l \} + \sin\{ n\pi (2l+x)/2l \}$$

$$+ \sin \{ n\pi (l + x) / 2l \} + (2/n\pi) \left[\cos\{n\pi(u - x) / 2l\}\right]_0^{-l}$$

$$+ \sin \{ n\pi (l - x) / 2l \} + (2/ n\pi)\left[\cos\{n\pi(u - x) / 2l\}\right]_0^{l}$$

$$+ \sin \{n\pi (2l - x) / 2l\} - \sin \{n\pi (l - x) / 2l \})$$

$$= (2l^2 / n\pi) (\sin \{ n\pi (2l + x) / 2l \} + (2/n\pi) [\cos \{n\pi (l + x) / 2l \}$$

$$- \cos (n\pi x / 2l) + \{\cos \{n\pi (l - x) / 2l \} - \cos (n\pi x/2l)]$$

$$+ \sin \{n\pi (2l - x) / 2l\})$$

$$= (2l^2/n\pi) [\sin \{n\pi (2l + x) / 2l\} + \sin \{n\pi (2l - x)/2l\}]$$

$$+ (2l/n\pi)^2 [\cos \{n\pi (l + x)/2l\} + \cos \{n\pi (l - x)/2l\} - 2. \cos (n\pi x/2l)]$$

$$= (4l^2 /n\pi) \sin n\pi. \cos (n\pi x / 2l)$$

$$+ (2l /n\pi)^2 \{ 2.\cos (n\pi/2). \cos (n\pi x /2l) - 2. \cos (n\pi x /2l) \}$$

$$= (8l^2 /n^2\pi^2) \{\cos (n\pi/2) - 1\} \cos (n\pi x /2l),$$

which is 0 when $n/2$ is even, $-(4l / n\pi)^2. \cos (n\pi x / 2l)$ when $n/2$ is odd, and $-(8l^2 /n^2 \pi^2). \cos (n\pi x /2l)$ for odd values of n. Hence, Eq. (6.2) becomes

$$f(x) = (3l /4) + (4l /\pi^2) \sum_{n=1}^{\infty} [\{\cos (n\pi/2) - 1\} \cos (n\pi x/2l)] / n^2$$

$$= (3l /4) - (4l /\pi^2) \{\cos (\pi x/2l) + (2/2^2) \cos (2\pi x/2l) + (1/3^2) \cos (3\pi x/2l)$$

$$+ (1/5^2) \cos (5\pi x/2l) + (2/6^2) \cos (6\pi x/2l) + \ldots\}$$

$$= (3l/4) - (4l/\pi^2) [\{\cos (\pi x/2l) + (1/3^2) \cos (3\pi x/2l) + (1/5^2) \cos (5\pi x/2l) + \ldots\}$$

$$+ 2\{ (1/2^2) \cos (\pi x/ l) + (1/6^2) \cos (3\pi x/ l) + (1/10^2) \cos (5\pi x/ l) +\ldots\}].//$$

Example 6.5. Find a series of *sines* of multiples of x which represent the function

$$f(x) = \begin{cases} x/2 \ (\text{when } 0 \le x \le \alpha), \\ \alpha/2 \ (\text{when } \alpha \le x \le \pi - \alpha), \\ (\pi - x)/2 \ (\text{when } \pi - \alpha \le x \le \pi), \end{cases}$$

on the interval $[0, \pi]$.

Solution. As in Example 5.3, in order to have a *sine* series expansion we construct a new (odd) function $g(x)$ on the extended interval $[-\pi, \pi]$. Thus, in view of Eq. (4.13), the Fourier coefficients are: $a_0 = a_n = 0$, and

$$b_n = (2/\pi)\int_0^{\pi} f(x)\sin nx\, dx = (2/\pi)\{\int_0^{\alpha} + \int_{\alpha}^{\pi-\alpha} + \int_{\pi-\alpha}^{\pi}\} f(x)\sin nx\, dx$$

$$= (1/\pi)\{\int_0^{\alpha} x.\sin nx\, dx + \alpha\int_{\alpha}^{\pi-\alpha} \sin nx\, dx + \int_{\pi-\alpha}^{\pi} (\pi - x).\sin nx\, dx$$

$$= (1/n\pi)\ \{ \left[-x.\cos nx\right]_0^{\alpha} + \int_0^{\alpha} \cos nx\, dx - \alpha \left[\cos nx\right]_{\alpha}^{\pi-\alpha}$$

$$- \left[(\pi - x)\cos nx\right]_{\pi-\alpha}^{\pi} - \int_{\pi-\alpha}^{\pi} \cos nx\, dx\ \}$$

$$= (1/n\pi)\ (-\alpha\cos n\alpha + (1/n)\left[\sin nx\right]_0^{\alpha} - \alpha\ \{\cos n(\pi - \alpha) - \cos n\alpha\}$$

$$+ \alpha.\cos n(\pi - \alpha) - (1/n)\left[\sin nx\right]_{\pi-\alpha}^{\pi}\ )$$

$$= (1/n^2\pi)\ \{\sin n\alpha + \sin n(\pi - \alpha)\}$$

$$= (2/n^2\pi)\sin(n\pi/2)\cos n(\pi/2 - \alpha) = 0, \text{ if } n \text{ is even.}$$

Therefore, Eq. (4.10b) assumes the form

$$f(x) = (2/\pi)\sum_{n=1}^{\infty} \{\sin(n\pi/2).\cos n(\pi/2 - \alpha).\sin nx\} / n^2$$

$$= (2/\pi)\{(\sin\alpha.\sin x)/1^2 + (\sin 3\alpha.\sin 3x)/3^2 + (\sin 5\alpha.\sin 5x)/5^2 + \ldots\}.//$$

CHAPTER 10

GREEN'S FUNCTION FOR LAPLACIAN OPERATOR

§ 1. *Laplace* Equation

The second order derivative

$$\nabla^2 = \nabla \cdot \nabla = \partial^2 / \partial x^2 + \partial^2 / \partial y^2 + \partial^2 / \partial z^2 \qquad (1.1)$$

defines the *Laplacian* operator. The function satisfying $\nabla^2 f = 0$ is called *harmonic* and the result is called the *Laplace* equation.

Example 1.1. The function $f = 1/r$ is harmonic, where $r = |\mathbf{r}|$ is the length of the position vector $\mathbf{r}$.

Solution. Differentiating $r^2 = x^2 + y^2 + z^2$ partially w.r.t. x we derive

$$r\, \partial r / \partial x = x \quad \Rightarrow \quad \partial r / \partial x = x/r.$$

Similarly,

$$\partial r / \partial y = y/r \quad \text{and} \quad \partial r / \partial z = z/r.$$

Therefore,

$$\nabla r = [\partial r / \partial x, \partial r / \partial y, \partial r / \partial z] = [x/r,\ y/r,\ z/r] = \mathbf{r} / r. \quad (1.2)$$

Also,

$$\nabla (1/r) = \text{grad}\,(1/r) = \{\hat{\mathbf{i}}\,(\partial/\partial x) + \hat{\mathbf{j}}\,(\partial/\partial y) + \hat{\mathbf{k}}\,(\partial/\partial z)\}\,(1/r)$$

$$= -(1/r^2)\,\nabla r = -(1/r^3)\,\mathbf{r},$$

by (1.2). Forming its dot product with ∇, we derive

$$\nabla^2 (1/r) = -\nabla.\, \{(1/r^3)\,\mathbf{r}\} = -\{\nabla (1/r^3)\}.\, \mathbf{r} - (1/r^3)\,(\nabla.\, \mathbf{r}),$$

by Eq. (2.3) of Chapter 1, Unit V. So, it is

$$= (3/r^4)\,(\nabla r).\, \mathbf{r} - 3/r^3, \quad \text{by Eq. (1.8) of Chapter 1, Unit V}$$

$$= (3/r^4)\, r - 3/r^3 = 0, \qquad \text{by (1.2) above,}$$

establishing the result. //

Example 1.2. Show that grad $r^m = m\, r^{m-2}\, \mathbf{r}$.

Solution. L.H.S. = $\{\hat{\mathbf{i}}\,(\partial/\partial x) + \hat{\mathbf{j}}\,(\partial/\partial y) + \hat{\mathbf{k}}\,(\partial/\partial z)\}\, r^m = m\, r^{m-1}\,(\nabla r)$

$$= m\, r^{m-1}\,(\mathbf{r}/r) = \text{R.H.S., by (1.2).} \; //$$

Example 1.3. Show that grad $f(r) \times \mathbf{r} = \mathbf{0}$.

Solution. $\text{grad}\, f(r) = \nabla f(r) = [\partial/\partial x, \partial/\partial y, \partial/\partial z]\, f(r)$

$$= f'(r)\,(\nabla r) = f'(r)\,(\mathbf{r}/r),$$

where $f'(r)$ is the total derivative of $f(r)$ w.r.t. r. Forming its cross product with $\mathbf{r}$ we get the desired result. //

Example 1.4. Show that grad $(\mathbf{r}\,.\,\mathbf{a}) = \mathbf{a}$.

Solution. $\mathbf{r}\,.\,\mathbf{a} = [x, y, z]\,.\,[a_1, a_2, a_3] = a_1 x + a_2 y + a_3 z.$

Hence,

$$\text{grad}\,(\mathbf{r}\,.\,\mathbf{a}) = \{\hat{\mathbf{i}}\,(\partial/\partial x) + \hat{\mathbf{j}}\,(\partial/\partial y) + \hat{\mathbf{k}}\,(\partial/\partial z)\}\,(a_1 x + a_2 y + a_3 z)$$

$$= a_1 \hat{\mathbf{i}} + a_2 \hat{\mathbf{j}} + a_3 \hat{\mathbf{k}} = \mathbf{a}\,. \; //$$

§ 2. Green's Theorem

Theorem 2.1. (*Green's theorem*) For two scalar functions f and g there hold

$$\int (\nabla f).(\nabla g)\, dv = \int \mathbf{n}\,.\,(f\,\nabla g)\, dS - \int (f\,\nabla^2 g)\, dv \qquad (2.1)$$

$$= \int \mathbf{n}\,.\,(g\,\nabla f)\, dS - \int (g\,\nabla^2 f)\, dv \qquad (2.2)$$

$\Rightarrow$

$$\int \mathbf{n}\,.\,(f\,\nabla g)\, dS - \int (f\nabla^2 g)\, dv = \int \mathbf{n}\,.\,(g\,\nabla f)\, dS - \int (g\,\nabla^2 f)\, dv, \quad (2.3)$$

where $\mathbf{n}$ is the unit normal vector to a closed surface (of area S) enclosing the volume v.

Proof. We recall the *Gauss divergence theorem*:

$$\int \mathbf{n}\,.\,\mathbf{F}\, dS = \int (\text{div}\, \mathbf{F})\, dv \qquad (2.4)$$

where the vector field $\mathbf{F}$ is defined over the surface S of volume v. Replacing $\mathbf{F}$ by $f\,\nabla g$ in (2.4) we conclude

$$\int \mathbf{n} \,.\, (f\ \nabla g)\, d\,S = \int \text{div}\,(f\ \nabla g)\, dv = \int \{(\nabla f).\,(\nabla g) + (f\ \nabla^2 g)\}\ dv,$$

by identity (2.3) of Chapter 1, Unit V. Arranging the terms there follows the identity (2.1).

Similarly, replacement of **F** by $g\,\nabla f$ in (2.4), yields

$$\int \mathbf{n} \,.\, (g\,\nabla f)\, d\,S\ = \int \text{div}\,(g\,\nabla f)\, dv = \int \{(\nabla\, g\,) \,.\, (\nabla f) + (g\ \nabla^2 f)\}\ dv.$$

Again, rearranging the terms, there follows the identity (2.2). Further, comparison of two results in identities (2.1) and (2.2) also yields the identity (2.3).

Corollary 2.1. There also results the identity

$$\int (\nabla f) \,.\, (\nabla\, g)\, dv\ =\ \int \mathbf{n} \,.\, (g\,\nabla f)\, dS \qquad (2.5)$$

for a harmonic function *f*.

Proof. In view of § 1, the Laplacian of a harmonic function vanishes: $\nabla^2 f = 0$. Hence, the identity (2.2) reduces to the form (2.5).

When both *f* and *g* are harmonic the identity (2.3) yields

$$\int \mathbf{n} \,.\, (f\ \nabla g\ -\ g\,\nabla f)\, d\,S\ =\ 0. \qquad (2.6)$$

Since

$$\mathbf{n} \,.\, \nabla g\ =\ [n_1, n_2, n_3] \,.\, [\partial g\,/\,\partial x,\ \partial g\,/\,\partial y,\ \partial g\,/\,\partial z]$$

$$= n_1\,(\partial g\,/\,\partial x) + n_2\,(\partial g\,/\,\partial y) + n_3\,(\partial g\,/\,\partial z)\ =\ dg/dn.$$

and similarly

$$\mathbf{n} \,.\, \nabla f\ =\ df\,/dn.$$

Hence, (2.6) further reduces to the form

$$\int \{f\,(dg\,/\,dn\ -\ g\,(d\,f/\,dn)\}\ d\,S\ =\ 0. \qquad (2.7)$$

CHAPTER 11

DIFFUSION EQUATION

§ 1. The Diffusion Equation

It is a partial differential equation (PDE). In physics, it describes the behavior of the collective motion of micro–particles in a material resulting from the random movement of each micro–particle. In mathematics, it is applicable in common to a subject relevant to the Markov process as well as in various other fields, such as the material sciences, information science, life science, social science, and so on. These subjects described by the diffusion equation are generally called *Brown problems.* The equation is usually written as:

$$\partial \varphi (\mathbf{r}, t) / \partial t = \nabla . \{ D (\phi, \mathbf{r}) \nabla \varphi (\mathbf{r}, t)\},$$

where ϕ $(\mathbf{r}, t)$ is the density of the diffusing material at location **r** and time t, D $(\phi, \mathbf{r})$ is the collective diffusion coefficient for density ϕ at location **r**, and ∇ represents the vector differential operator. If the diffusion coefficient depends on the density then the equation is nonlinear, otherwise it is linear.

More generally, when D (not to be confused with derivation) is a symmetric positive definite matrix, the equation describes anisotropic diffusion, which is written (for 3–dimensional case) as:

$$\partial \varphi (\mathbf{r}, t) / \partial t = \sum_{i, j=1}^{n} \frac{\partial}{\partial x^i} \left\{ D_{ij} (\varphi, \mathbf{r}) \frac{\partial \varphi (\mathbf{r}, t)}{\partial x^j} \right\}.$$

If D is constant, then the equation reduces to the following linear differential equation:

$$\partial \varphi (\mathbf{r}, t) / \partial t = D \nabla^2 \varphi (\mathbf{r}, t)$$

also called the *heat equation.*

§ 2. Heat Equation

It is also an important PDE governing the temperature u in a body in space. We obtain this model of temperature distribution under the following assumptions.

Physical Assumptions

(i) The *specific heat* σ and the *density* ρ of the material of the body are constant. No heat is produced or it disappears in the body.

(ii) Experiments show that, in a body, heat flows in the direction of decreasing temperature, and the rate of flow is proportional to the gradient of the temperature; that is, the velocity **v** of the heat flow in the body is of the form

$$\mathbf{v} = -K \operatorname{grad} u, \tag{2.1}$$

where $u(x, y, z, t)$ is the temperature at a point and time t.

(iii) The *thermal conductivity* K is constant, as is the case for homogeneous material and non-extreme temperatures.

Under these assumptions we can model heat flow as follows.

Let R be a region in the body bounded by a surface S with outer unit normal vector **n** such that the (Gauss) divergence theorem applies. Then $\mathbf{v} \cdot \mathbf{n}$ is the component of **v** in the direction of **n**. Hence, $|\mathbf{v} \cdot \mathbf{n}\, \Delta A|$ is the amount of heat *leaving* R (if $\mathbf{v} \cdot \mathbf{n} > 0$ at some point P) or *entering* R (if $\mathbf{v} \cdot \mathbf{n} < 0$ at P) per unit time at some point P of S through a small portion ΔS of S of area ΔA. Hence the total amount of heat that flows across S from R is given by the surface integral $\iint_S \mathbf{v} \cdot \mathbf{n}\, dA$. Using Gauss divergence theorem, we now convert our surface integral into a volume integral over the region R. Because of (2.1) this gives

$$\iint_S \mathbf{v}.\mathbf{n}\, dA = -K \iint_S (\operatorname{grad} u) \cdot \mathbf{n}\, dA$$

$$= -K \iiint_R \operatorname{div}(\operatorname{grad} u)\, dx\, dy\, dz = -K \iiint_R (\nabla^2 u)\, dx\, dy\, dz. \tag{2.2}$$

On the other hand, the total amount of heat in R can be written as $H = \iiint_R (\sigma\rho\, u)\, dx\, dy\, dz$ with σ and ρ as before. Hence the time rate of decrease of H is $-\partial H/\partial t = -\iiint_R (\sigma\rho)(\partial u/\partial t)\, dx\, dy\, dz$. This must be equal to the amount of heat leaving R because no heat is produced or disappears in the body. From (2.2), we thus obtain

$$-\partial H/\partial t = -\iiint_R (\sigma\rho)(\partial u/\partial t)\,dx\,dy\,dz = -K\iiint_R (\nabla^2 u)\,dx\,dy\,dz,$$

or (dividing by $-\sigma\rho$ and putting $c^2 = K/\sigma\rho$)

$$\iiint_R (\partial u/\partial t - c^2 (\nabla^2 u)\,dx\,dy\,dz = 0.$$

Since this holds for any region R in the body, the integrand (if continuous) must be zero everywhere. That is,

$$\partial u/\partial t = c^2 (\nabla^2 u). \qquad (2.3)$$

This is the **heat equation** (the fundamental PDE) modeling heat flow. It gives the temperature $u(x, y, z, t)$ in a body of homogeneous material in space. The constant c^2 is the *thermal diffusivity*. K is the *thermal conductivity*, σ the *specific heat*, and ρ the *density* of the material of the body. $\nabla^2 u$ is the Laplacian of u and, with respect to the Cartesian coordinates x, y, z. The heat equation is also called the **diffusion equation** because it also models chemical diffusion processes of one substance or gas into another.

CHAPTER 12

GREEN'S FUNCTION FOR THE HELMHOLTZ OPERATOR

§ 1. Introduction

The non–homogeneous Helmholtz differential equation is

$$\nabla^2 \psi(r) + k^2 \psi(r) = \rho(r), \tag{1.1}$$

where the Helmholtz operator is defined as $\tilde{L} \equiv \nabla^2 + k^2$. The Green's function is then defined by

$$(\nabla^2 + k^2) G(r_1, r_2) = \delta^3 (r_1 - r_2). \tag{1.2}$$

Defining the basic function φ_n as the solutions to the homogeneous Helmholtz differential eqn.

$$\nabla^2 \varphi_n(r) + k^2 \varphi_n(r) = 0. \tag{1.3}$$

The Green's function can then be expanded in terms of the φ_n,

$$G(r_1, r_2) = \sum_{n=0}^{\infty} a_n(r_2) . \varphi_n(r_1), \tag{1.4}$$

and the *delta function* as

$$\delta^3 (r_1 - r_2) = \sum_{n=0}^{\infty} \varphi_n(r_1) . \varphi_n(r_2). \tag{1.5}$$

Plugging (1.4) and (1.5) into (1.2) gives

$$\nabla^2 \left\{ \sum_{n=0}^{\infty} a_n(r_2) . \varphi_n(r_1) \right\} + k^2 \left\{ \sum_{n=0}^{\infty} a_n(r_2) . \varphi_n(r_1) \right\}$$
$$= \sum_{n=0}^{\infty} \varphi_n(r_1) . \varphi_n(r_2). \tag{1.6}$$

Putting from Eq. (1.3), it gives

$$-\left\{\sum_{n=0}^{\infty} a_n(r_2).k_n^{\ 2}.\varphi_n(r_1)\right\} + k^2\left\{\sum_{n=0}^{\infty} a_n(r_2).\varphi_n(r_1)\right\}$$

$$= \sum_{n=0}^{\infty} \varphi_n(r_1).\varphi_n(r_2), \quad (1.7)$$

or,

$$\left\{\sum_{n=0}^{\infty} a_n(r_2).\varphi_n(r_1).(k^2 - k_n^{\ 2})\right\} = \sum_{n=0}^{\infty} \varphi_n(r_1).\varphi_n(r_2). \quad (1.8)$$

This equation must hold true for each n, so

$$a_n(r_2).\varphi_n(r_1).(k^2 - k_n^{\ 2}) = \varphi_n(r_1).\varphi_n(r_2). \quad (1.9)$$

$\Rightarrow$

$$a_n(r_2) = \varphi_n(r_2)/(k^2 - k_n^{\ 2}),$$

and (1.4) can be written

$$G(r_1, r_2) = \sum_{n=0}^{\infty} \varphi_n(r_1).\varphi_n(r_2)/(k^2 - k_n^{\ 2}).$$

The general solution to (1.1) is therefore

$$\psi(r_1) = \int G(r_1, r_2).\ \rho(r_2).d^3 r_2$$

$$= \sum_{n=0}^{\infty} \int\left\{\varphi_n(r_1).\varphi_n(r_2).\ \rho(r_2)/(k^2 - k_n^{\ 2})\right\}.d^3 r_2. \quad (1.12)$$

CHAPTER 13

PARTIAL DERIVATION AND WAVE EQUATION

§ 1. Introduction

An equation involving *one* or *more* partial derivatives of an unknown function of *two* or *more* variables is called a *partial differential equation*. It is said to be *linear* if it is of the *first* degree in the *dependent variable* and its *partial derivatives.*

The following are some important *linear* partial differential equations of the *second* order (wherein c is a constant and t the time parameter):

Eqn.	Heat Equation	Wave Equation	Laplace Equation	Poisson Equation
1-dim	$c^2\,\partial^2 u / \partial x^2 = \partial u/\partial t$	$c^2\,\partial^2 u / \partial x^2 = \partial^2 u / \partial t^2$		
2-dim		$c^2 (\partial^2 u / \partial x^2 + \partial^2 u / \partial y^2) = \partial^2 u / \partial t^2$	$\partial^2 u / \partial x^2 + \partial^2 u / \partial y^2 = 0$	$\partial^2 u / \partial x^2 + \partial^2 u / \partial y^2 = f(x, y)$
3-dim			$\nabla^2 u \equiv \partial^2 u / \partial x^2 + \partial^2 u / \partial y^2 + \partial^2 u / \partial z^2 = 0$	

A *solution* of a partial differential equation in some region $\Re$ of the space of the independent variable is a function which has all the partial derivatives appearing in the equation in some domain containing $\Re$ and satisfies the equation everywhere in $\Re$.

Note 1.1. We note that all $u = x^2 - y^2$, $u = e^x. \cos y$, $u = \ln (x^2 + y^2)$ which are entirely different from each other are solutions of two-dimensional Laplace equation.

The *unique solution* can be determined from the additional information given concerning the variables of the equation. We call the additional information as *initial conditions* or more generally the *boundary conditions* because they do not always refer to *zero* values of the independent variables.

Example 1.1. Solve the differential equation

$$\partial^2 u / \partial x^2 = 12\, x^2 (t + 1) \tag{1.1}$$

given that

$$u\,(0) = \cos 2t, \quad \text{and} \quad (\partial u/\partial x)_{x=0} = \sin t. \tag{1.2}$$

Solution. We note that x and t are independent variables and u is a function of both x and t. Integrating the differential equation (1.1) partially w.r.t. x, we get

$$\partial u / \partial x = 4\, x^3 . (t + 1) + f(t), \tag{1.3}$$

where $f(t)$ is a function of t alone and is independent of x. Applying the second initial condition, we evaluate $f(t) = \sin t$. Integrating Eq. (1.3) again partially w.r.t. x, we obtain

$$u = x^4 . (t + 1) + x . \sin t + g\,(t),$$

where $g\,(t)$ is also independent of x. Now applying the first initial condition, we also evaluate $g\,(t) = \cos 2t$. Hence, the desired solution is

$$u = x^4 . (t + 1) + x . \sin t + \cos 2t. \ //$$

Example 1.2. Solve the differential equation

$$\partial^2 u / \partial x\, \partial y = \sin (x + y), \tag{1.4}$$

given that

$$u\,(0) = (y - 1)^2 \quad \text{and} \quad (\partial u/\partial x)_{y=0} = 1. \tag{1.5}$$

Solution. Here, x and y are independent variables and u is a function of both x and y. Integrating Eq. (1.4) partially w.r.t. y, we get

$$\partial u / \partial x = -\cos (x + y) + f(x), \tag{1.6a}$$

where $f(x)$ is independent of y. Applying the second initial condition, we evaluate $f(x) = 1 + \cos x$. Hence, Eq. (1.6a) reduces to

$$\partial u / \partial x = -\cos (x + y) + 1 + \cos x. \tag{1.6b}$$

Integrating it again partially w.r.t. x, we obtain

$$u = -\sin(x+y) + x + \sin x + g(y), \tag{1.7}$$

where $g(y)$ is independent of x. Now applying the first initial condition, we evaluate $g(y) = (y-1)^2 + \sin y$. Hence, Eq. (1.7) gives the desired solution

$$u = -\sin(x+y) + x + \sin x + (y-1)^2 + \sin y. \ //$$

§ 2. Separation of Variables

For a homogeneous linear partial differential equation, it is sometimes possible to find particular solution in the form of a product of two independent functions. Let us consider a function $u(x, y)$ of two independent variables x and y in a product form:

$$u = X(x).\,Y(y), \tag{2.1}$$

where X is a function of x only and Y a function of y only. As immediate consequences of this we have

$$\left.\begin{aligned} \partial u/\partial x &= X'(x).Y(y), & \partial u/\partial y &= X(x).Y'(y), \\ \partial^2 u/\partial x^2 &= X''(x).Y(y), & \partial^2 u/\partial y^2 &= X(x).Y''(y), \end{aligned}\right\} \tag{2.2}$$

where the primes denote ordinary differentiation w.r.t. the respective variables. This method is applicable for the solution of partial differential equations where variables are separable.

Example 2.1. Find the product solutions of the partial differential equation

$$\partial^2 u/\partial x^2 = 4\,(\partial u/\partial y). \tag{2.3}$$

Solution. Let Eq. (2.1) provide a product solution then, in view of Eq. (2.2), Eq. (2.3) reduces to

$$X''.Y = 4\,X.Y' \quad \Rightarrow \quad X''/4X = Y'/Y. \tag{2.4}$$

Since the LHS of Eq. (2.4) is independent of y whereas the RHS is independent of x. But, the two sides being identically equal, we conclude that expressions on either sides must be constant, say k. Often, it is convenient to write this real constant as either λ^2 or $-\lambda^2$. In the following, we consider the different cases.

Case 2.1. Taking $k = \lambda^2 > 0$, Eq. (2.4) leads to

$$X'' - 4\lambda^2 . X = 0 \quad (2.5); \qquad Y' - \lambda^2 . Y = 0. \quad (2.6)$$

These equations have solutions (cf. Eq. (2.4.10b))

$$X = A_1 . \cosh(2\lambda x) + B_1 . \sinh(2\lambda x), \tag{2.7}$$

(respectively)

$$Y = C_1 . \exp(\lambda^2 y). \tag{2.8}$$

Thus, a particular solution of Eq. (2.3), by Eq. (2.1), is

$$u = (A_1 . \cosh 2\lambda x + B_1 . \sinh 2\lambda x)\{C_1 . \exp(\lambda^2 y)\}$$

$$= (A_2 . \cosh 2\lambda x + B_2 . \sinh 2\lambda x) . \exp(\lambda^2 y), \tag{2.9}$$

where we have put $A_2 = A_1 . C_1$ and $B_2 = B_1 . C_1$.

Case 2.2. Taking $k = -\lambda^2 < 0$, Eq. (2.4) leads to

$$X'' + 4\lambda^2 . X = 0, \tag{2.10}$$

and

$$Y' + \lambda^2 . Y = 0. \tag{2.11}$$

These equations have solutions (cf. Eq. (2.4.8b))

$$X = A_3 . \cos 2\lambda x + B_3 . \sin 2\lambda x, \tag{2.12}$$

(respectively)

$$Y = C_3 . \exp(-\lambda^2 y). \tag{2.13}$$

Thus, a particular solution of Eq. (2.3) is

$$u = (A_3 . \cos 2\lambda x + B_3 . \sin 2\lambda x)\{C_3 . \exp(-\lambda^2 y)\}$$

$$= (A_4 . \cos 2\lambda x + B_4 . \sin 2\lambda x) . \exp(-\lambda^2 y) \tag{2.14}$$

where we have put $A_4 = A_3 . C_3$ and $B_4 = B_3 . C_3$.

Case 2.3. Taking $k = 0$, Eq. (2.4) leads to $X'' = 0$ and $Y' = 0$ having their solutions $X = A_5.x + B_5$ and $Y = C_5$ respectively. Thus, a particular solution of Eq. (2.3) is

$$u = (A_5.x + B_5).\ C_5 = A_6.x + B_6, \qquad (2.15)$$

where $A_6 = A_5.C_5$ and $B_6 = B_5.C_5$. //

Definition 2.1. (Superposition principle) If $u_1, u_2, u_3, \ldots, u_n$, are some solutions of a linear partial differential equation, then their linear combination

$$U = c_1.\ u_1 + c_2.\ u_2 + c_3.\ u_3 + \ldots + c_n.\ u_n\ , \qquad (2.16)$$

where c_i's, $i = 1, 2, 3, \ldots, n$, are constants, is also a solution.

Example 2.2. Solve the following partial differential equations:

(i) $\partial u / \partial x + y = 0$, **(ii)** $\partial u / \partial x + u = e^y$, **(iii)** $\partial^2 u / \partial x\ \partial y = 1$.

Solution. (i) Integrating the differential equation partially w.r.t. x, we get

$$u + x y = f(y),$$

where $f(y)$ is an arbitrary constant of integration with respect to x and is a function of y alone.

(ii) This is a linear (in the dependent variable u and its partial derivative $\partial u/\partial x$) partial differential equation of first order. The integrating factor is $e^{\int 1\, dx} = e^x$. Therefore, its partial integration w.r.t. x (treating the other independent variable y constant) gives

$$u.\ e^x = e^y. \int e^x.\ dx + f(y) = e^y.\ e^x + f(y) \Rightarrow u = e^y + e^{-x}.\ f(y),$$

where $f(y)$ is independent of x.

(iii) Integrating the differential equation partially w.r.t. y (treating the other independent variable x constant), we get

$$\partial u / \partial x = y + f(x),$$

where $f(x)$ is an arbitrary constant of integration (with respect to y) and is a function of x alone. Integrating it further partially w.r.t. x (now treating y constant), we get

$$u = xy + \int f(x).\, dx + g(y),$$

where $g(y)$ is an arbitrary constant of integration w.r.t. x and is a function of y alone. //

§ 3. One-dimensional Wave Equation

Consider a perfectly flexible elastic string of (natural) length l stretched between two points O ($x = 0$) and A ($x = l$) with uniform tension T. Let the string be displaced slightly from its initial position of rest and released while the end points remain fixed. The string will then vibrate and the vibrations are governed by the one dimensional wave equation written as

$$c^2(\partial^2 u / \partial x^2) = \partial^2 u / \partial t^2. \tag{3.1}$$

Assuming the initial conditions

$$x = 0, \qquad u(0, t) = 0, \tag{3.2a}$$

$$x = l, \qquad u(l, t) = 0, \tag{3.2b}$$

for all values of t and taking the initial deflection and initial velocity as the following

$$u(x, 0) = f(x) \quad (3.3); \qquad (\partial u / \partial t)_{t=0} = g(x), \tag{3.4}$$

we need to find a solution of Eq. (3.1). This is carried out in the following three steps.

Step 1. Analogous to Eq. (2.1), let

$$u = X(x).\, T(t) \tag{3.5}$$

be a trial solution of Eq. (3.1) so that, in analogy with Eq. (2.2), there hold the relations

$$\left.\begin{aligned} u_x \equiv \partial u / \partial x = X'(x).\, T(t), \quad u_t \equiv \partial u / \partial t = X(x).\, T'(t) \\ u_{xx} \equiv \partial^2 u / \partial x^2 = X''(x).T(t), \quad u_{tt} \equiv \partial^2 u / \partial t^2 = X(x).T''(t) \end{aligned}\right\} \tag{3.6}$$

reducing Eq. (3.1) to the form

$$c^2 X''.T = X.T'' \quad \Rightarrow \quad X''/X = T''/c^2 T. \qquad (3.7)$$

The expressions on the two sides of Eq. (3.7) are functions of independent) variables x and t respectively so either of them must be constant, say k. Thus, the equations (3.7) lead to

$$X'' - k.X = 0 \quad (3.8); \quad \text{and} \quad T'' - c^2.k.T = 0. \qquad (3.9)$$

Step 2. We now seek the solutions of these differential equations so that Eq. (3.5) satisfies the initial conditions given by Eq. (3.2) for every value of t:

$$u(0, t) = X(0).T(t) = 0, \qquad (3.10a)$$

$$u(l, t) = X(l).\,T(t) = 0. \qquad (3.10b)$$

We discuss both vanishing and non-vanishing choices of the function T: $T = 0$ gives a trivial solution of Eq. (3.1): $u \equiv X.T = 0$, which is of no interest to us. Also, when $T \neq 0$, from Eq. (3.10), there result

$$X(0) = 0, \text{ as well as } X(l) = 0. \qquad (3.11)$$

In the following, we will discuss various possibilities for the constant k.

(i) If $k = 0$, the differential equation (3.8) has a general solution $X = a.x + b$; which, for the initial conditions given by Eq. (3.11), determines

$$0 = a.0 + b \Rightarrow b = 0; \quad \text{and} \quad 0 = X(l) = a.l + 0 \Rightarrow a = 0, \quad \text{as} \quad l \neq 0.$$

Hence, the solution of Eq. (3.8) becomes $X = 0 \Rightarrow u \equiv X.\,T = 0$ (as before for $T = 0$).

(ii) If $k > 0$, say ρ^2, then Eq. (3.8) assumes the form $X'' - \rho^2 X = 0$ having a solution

$$X = Ae^{\rho x} + Be^{-\rho x}. \qquad (3.12)$$

Applying the initial conditions given by Eq. (3.11), we evaluate the constants A and B:

$$X(0) = 0 = A + B \quad \text{and} \quad X(l) = 0 = Ae^{\rho l} + Be^{-\rho l}.$$

Eliminating B from these we get

$$A.(e^{\rho l} - e^{-\rho l}) = 0 \quad \Rightarrow \quad A = 0, \text{ for } e^{\rho l} - e^{-\rho l} \neq 0.$$

Hence, $B = -A = 0$. Thus, Eq. (3.12) again yields $X = 0 \Rightarrow u = 0$ as before.

(iii) When $k < 0$, say $-\rho^2$, then Eq. (3.8) reduces to $X'' + \rho^2 X = 0$ having a general solution

$$X = A \cos \rho x + B \sin \rho x. \tag{3.13}$$

Applying the initial conditions given by Eq. (3.11), we derive $X(0) = 0 = A$, and $X(l) = B \sin \rho l = 0$. The latter relation gives two alternatives: either $B = 0$, or

$$\sin \rho l = 0. \tag{3.14}$$

The choice $B = 0$ (together with $A = 0$ as seen above) again reduces Eq. (3.13) to $X = 0$ implying $u = 0$, which is trivial. But, the latter alternative Eq. (3.14) yields $\rho l = 0$, or $n\pi$ (n being an integer). Since $l \neq 0$ and ρ (the mass per unit length of the string) too cannot be zero leaving the possibility

$$\rho = n\pi / l. \tag{3.15}$$

Setting $B = 1$ (for simplicity), we obtain infinitely large number of solutions of Eq. (3.8):

$$X \equiv X_n(x) = \sin(n\pi x / l), \; n = 1, 2, 3, \ldots \tag{3.16}$$

For Eq. (3.15), $k \equiv -\rho^2 = -(n\pi / l)^2$, the equation (3.9) takes the form

$$T'' + \lambda_n^2 . T = 0, \tag{3.17}$$

where

$$\lambda_n \equiv c\,\rho = c\, n\pi / l. \tag{3.18}$$

A general solution of Eq. (3.17) is

$$T_n(t) = A_n . \cos(\lambda_n t) + B_n . \sin(\lambda_n t). \tag{3.19}$$

Thus, Eqs. (3.16) and (3.19) determine solutions of Eq. (3.1):

$$u_n(x, t) \equiv X_n(x).T_n(t) = \{A_n.\cos(\lambda_n t) + B_n.\sin(\lambda_n t)\} \sin(n\pi x / l). \tag{3.20}$$

Definition 3.1. Above solutions of the wave equation (3.1) are called the *eigen functions* and the values λ_n given by Eq. (3.18) as the *eigen values* of the vibrating string.

Step 3. From Eq. (3.20), we derive

$$u_n(x, 0) = A_n. \sin(n\pi x / l) = f(x), \tag{3.21}$$

for Eq. (3.3); and

$$\partial u_n / \partial t = \{-A_n. \sin(\lambda_n t) + B_n. \cos(\lambda_n t)\}.\lambda_n. \sin(n\pi x/ l)$$

$\Rightarrow$

$$(\partial u_n / \partial t)_{t=0} = B_n \lambda_n \sin(n\pi x/ l) = g(x), \tag{3.22}$$

for Eq. (3.4). Comparing Eqs. (3.21) and (3.22), we note that the functions $f(x)$ and $g(x)$ are not independent and are connected by a linear relation

$$(B_n \lambda_n / A_n)\, f(x) = g(x).$$

As Eq. (3.1) is linear and homogeneous we cannot accept Eq.(3.20) as a solution. Instead, we consider the infinite series

$$u(x, t) \equiv \sum_{n=1}^{\infty} u_n(x, t)$$

$$= \sum_{n=1}^{\infty} \{A_n. \cos(\lambda_n t) + B_n. \sin(\lambda_n t)\} \sin(n\pi x / l). \tag{3.23}$$

At $t = 0$, we, therefore, have

$$u(x,0) \equiv f(x) = \sum_{n=1}^{\infty} \{A_n. \sin(n\pi x / l). \tag{3.24}$$

This gives an expansion of $f(x)$ in a *sine* series. Hence, in view of Eq. (9.5.16), we get

$$A_n = (2/l)\int_{x=0}^{l} f(x) \sin(n\pi x / l), dx, n = 1, 2, 3, \ldots \tag{3.25}$$

Also, in view of Eq. (3.22), we have

$$(\partial u / \partial t)_{t=0} \equiv \sum_{n=1}^{\infty} (\partial u_n/\partial t)_{t=0} = g(x) = \sum_{n=1}^{\infty} B_n \lambda_n \sin(n\pi x/l). \tag{3.26}$$

This, in analogy with Eq. (3.25), determines

$$B_n \lambda_n = (2/l) \int_{x=0}^{l} g(x).\ \sin(n\pi x/l).\ dx,$$

or, for Eq. (3.18),

$$B_n = (2/cn\pi) \int_{x=0}^{l} g(x) \sin(n\pi x/l).\ dx,\ n = 1, 2, 3, \ldots \quad (3.27)$$

Thus, Eq. (3.23) gives a solution of the wave equation (3.1) where the constants A_n, B_n are determined by Eqs. (3.25) and (3.27).

3.1. Particular case: When the string is released from rest $g(x) = 0$ for every x in the interval $0 \le x \le l$, it follows from Eq. (3.27) that $B_n = 0$ reducing the solution (3.23) to

$$u(x, t) = \sum_{n=1}^{\infty} A_n.\ \cos(\lambda_n t).\ \sin(n\pi x/l). \quad (3.28)$$

Example 3.1. Let a vibrating string of length 30 cms. satisfies the wave equation

$$4(\partial^2 u/\partial x^2) = \partial^2 u/\partial t^2, \quad 0 < x < 30,\ 0 < t. \quad (3.29)$$

Let the ends of the string be fixed and it is set in motion with zero initial velocity from the initial position:

$$u(x, 0) = f(x) = \begin{cases} x/10, & 0 \le x \le 10, \\ (30 - x)/20, & 10 \le x \le 30. \end{cases} \quad (3.30)$$

Find the displacement $u(x, t)$ of the string.

Solution. A solution of the wave equation is given by equation (3.28). For $c = 2$, $l = 30$ determining $\lambda_n = cn\pi/l = 2n\pi/30 = n\pi/15$, the equation (3.28) becomes

$$u(x, t) = \sum_{n=1}^{\infty} A_n \cos(n\pi t/15).\ \sin(n\pi x/30). \quad (3.31)$$

Also, Eq. (3.25) determines A_n:

$$A_n = (2/30). \int_{x=0}^{30} f(x).\ \sin(n\pi x/30).\ dx$$

$$= (1/15)\left\{\int_{x=0}^{10} (x/10).\sin(n\pi x/30).\,dx\right.$$

$$\left. + \int_{x=10}^{30} \{(30 - x)/20\}.\sin(n\pi x/30).\,dx\right\}$$

$$= (1/150)\int_{x=0}^{10} x.\sin(n\pi x/30).\,dx + (1/10)\int_{x=10}^{30} \sin(n\pi x/30).dx$$

$$- (1/300)\int_{x=10}^{30} x.\sin(n\pi x/30).dx$$

$$= (1/5n\pi)\left\{\left[-x.\cos(n\pi x/30)\right]_0^{10} + \int_{x=0}^{10} \cos(n\pi x/30).\,dx\right\}$$

$$+ (3/n\pi)\left[-\cos(n\pi x/30)\right]_{10}^{30}$$

$$+ (1/10n\pi)\left\{\left[x.\cos(n\pi x/30)\right]_{10}^{30} - \int_{x=10}^{30} \cos(n\pi x/30).\,dx\right\}$$

$$= (-2/n\pi)\cos(n\pi/3) + (6/n^2\pi^2)\left[\sin(n\pi x/30)\right]_0^{10}$$

$$+ (3/n\pi)\{-\cos n\pi + \cos(n\pi/3)\} + (1/n\pi)\{3\cos n\pi - \cos(n\pi/3)\}$$

$$- (3/n^2\pi^2)\left[\sin(n\pi x/30)\right]_{10}^{30}$$

$$= (6/n^2\pi^2)\sin(n\pi/3) - (3/n^2\pi^2)\{\sin n\pi - \sin(n\pi/3)\}$$

$$= (9/n^2\pi^2)\sin(n\pi/3). \;//$$

CHAPTER 14

THE ROTATION GROUP

§ 1. Translation

When a figure moves through a given distance in a prescribed direction without turning, its movement is called a *translation*. Thus, a translation is a transformation of a plane that shifts every point in the plane the same distance in the same direction. If A is an object with A′ as its image under a translation T, we then write

$$\text{T: A} \rightarrow \text{A}', \qquad \text{or equivalently} \qquad \text{T (A)} = \text{A}'.$$

Under translation if one point moves then all points move and no point remains fixed.

Example 1.1. Under translation the image of a circle (respectively angle) is a circle (respectively angle).

1.1. Translation in the coordinate plane

If a translation has tail at origin and its head at the point (a, b), then the image of a point (x, y) under the translation is $(x + a, y + b)$. Similarly, if a translation has tail at (h, k) and the head at (a, b), then the image of (x, y) is $(x + a - h, y + b - k)$.

Example 1.2. A translation carries the origin onto the point $(-3, 7)$. Then, the image of $(-8, -4)$ is $(-11, 3)$.

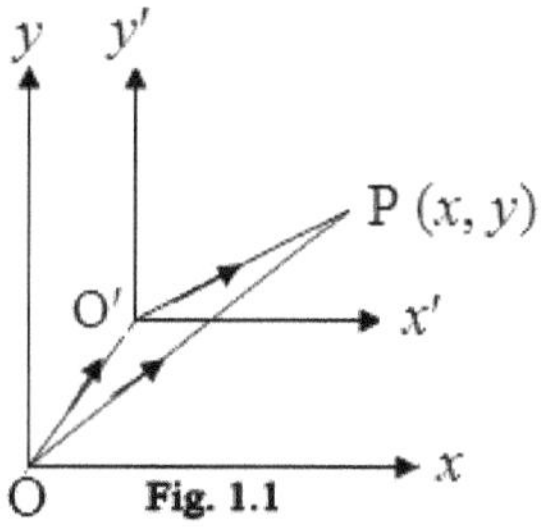

Fig. 1.1

1.2. Representation of a translation in matrix form

Let P be any point in a plane with Cartesian coordinates (x, y) with respect to a system of coordinates, say S, having origin at O and the rectangular coordinate axes Ox and Oy. If the origin is shifted to a point O′ with coordinates (a, b) with respect to S and the coordinate axes Ox and Oy are translated to O′ x' and O′ y' respectively then the translation causes the transformation of vectors according to

$$(x, y) \equiv \overrightarrow{\text{OP}} = \overrightarrow{\text{OO}'} + \overrightarrow{\text{O}'\text{P}} = (a, b) + (x', y'). \qquad (1.1)$$

Thus,

$$x = x' + a, \qquad y = y' + b. \tag{1.2}$$

These transformations are also expressed in the reverse order:

$$x' = x - a, \qquad y' = y - b \tag{1.3}$$

describing the translation of the coordinate axes. Denoting the homogeneous Cartesian coordinates [cf. Ref. 9, Chap. 1] of P by $(x, y, 1)$ and $(x', y', 1)$ with respect to two systems of coordinates: S and S′ above transformations can be expressed in terms of a matrix equation

$$\begin{bmatrix} x' \\ y' \\ 1 \end{bmatrix} = \begin{bmatrix} 1 & 0 & -a \\ 0 & 1 & -b \\ 0 & 0 & 1 \end{bmatrix} \cdot \begin{bmatrix} x \\ y \\ 1 \end{bmatrix} \quad (1.4); \qquad \text{where} \quad T = \begin{bmatrix} 1 & 0 & -a \\ 0 & 1 & -b \\ 0 & 0 & 1 \end{bmatrix}$$

is called the *translation matrix*. Every translation matrix for a plane is of this form only and has its determinant of value 1.

§ 2. Reflection

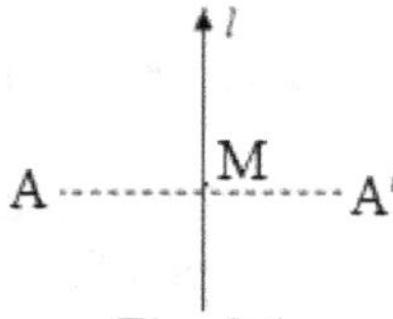

Fig. 2.1

A reflection through a line l is a transformation of the plane such that:

(i) If $A \notin l$, the image of A is A′, provided l is the perpendicular bisector of AA′; thus causing AM = MA′;

(ii) If A ε l, the image of A is A itself.

The line l is called the *axis of reflection* or the *line of reflection*. A reflection always preserves the distance. Thus, a line segment (under a reflection) is a congruent line segment. Reflections of different line segments AB in a line l are caused in the following ways:

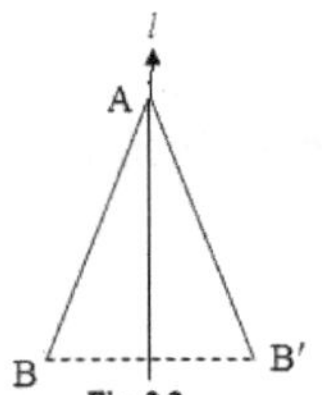

Fig. 2.2

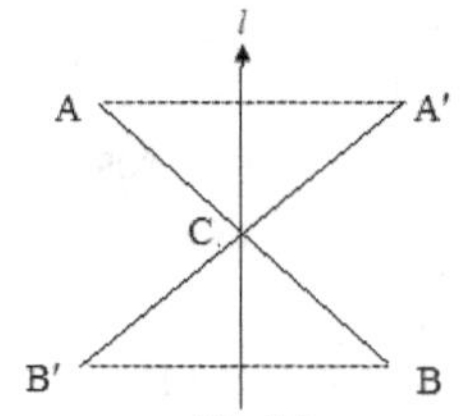

Fig. 2.3

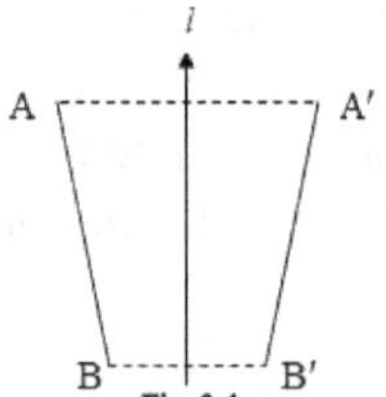

Fig. 2.4

Example 2.1. A circle is reflected into itself about a line through its centre. Such a line is called the *line* (or the *axis*) *of symmetry*.

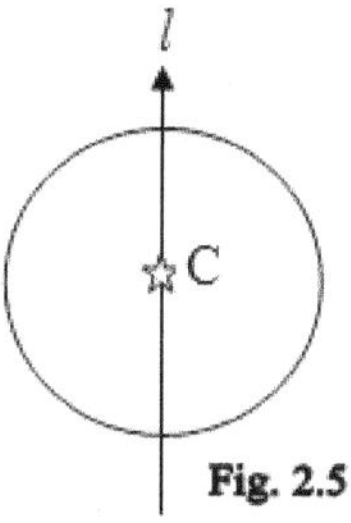

Fig. 2.5

Note 1.1. A figure is called *symmetric* if it has at least one line of symmetry.

Example 2.2. Under reflection through x–axis a point P (x, y) reflects onto the point P′ $(x, -y)$. Denoting the reflection through x–axis by R_x we, thus, have

$$R_x(x, y) = (x, -y). // \qquad (2.1)$$

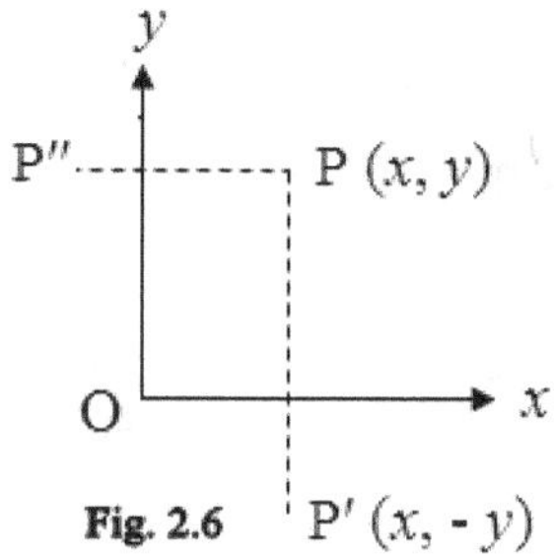

Fig. 2.6

Example 2.3. Under reflection through y–axis a point P (x, y) reflects onto the point P″ $(-x, y)$. Denoting the reflection through y–axis by R_y we, thus, have

$$R_y(x, y) = (-x, y). // \qquad (2.2)$$

Note 2.1. Reflections through the coordinate axes Ox and Oy can also be represented by matrices

$$\begin{bmatrix} 1 & 0 & 0 \\ 0 & -1 & 0 \\ 0 & 0 & 1 \end{bmatrix} \text{ and } \begin{bmatrix} -1 & 0 & 0 \\ 0 & 1 & 0 \\ 0 & 0 & 1 \end{bmatrix}$$

respectively. Therefore, (2.1), (2.2) are rewritten in matrix forms:

$$\begin{bmatrix} x' \\ y' \\ 1 \end{bmatrix} = \begin{bmatrix} 1 & 0 & 0 \\ 0 & -1 & 0 \\ 0 & 0 & 1 \end{bmatrix} \cdot \begin{bmatrix} x \\ y \\ 1 \end{bmatrix} = \begin{bmatrix} x \\ -y \\ 1 \end{bmatrix}, \qquad (2.1a)$$

$$\begin{bmatrix} x' \\ y' \\ 1 \end{bmatrix} = \begin{bmatrix} -1 & 0 & 0 \\ 0 & 1 & 0 \\ 0 & 0 & 1 \end{bmatrix} \cdot \begin{bmatrix} x \\ y \\ 1 \end{bmatrix} = \begin{bmatrix} -x \\ y \\ 1 \end{bmatrix}. \qquad (2.2a)$$

Example 2.4. Under reflection through the line $y = x$, a point P (x, y) reflects onto the point P_2 (y, x):

$$R_{y=x}(x, y) = (y, x). \qquad (2.3)$$

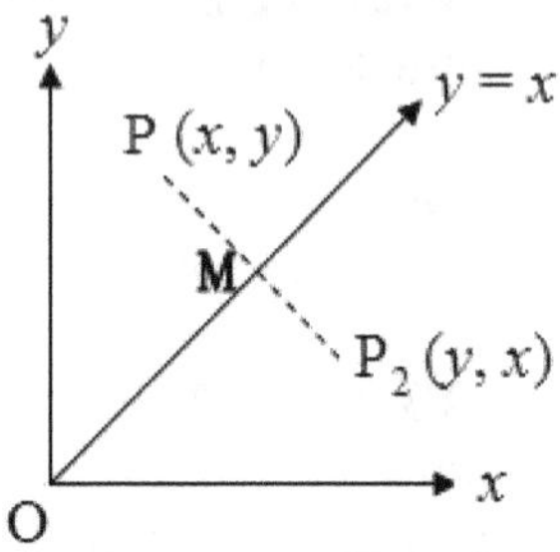

Solution. Let the coordinates of P and P_2 be (x_1, y_1) and (x_2, y_2) respectively. As per definition of the reflection, the line $y = x$ is the perpendicular bisector of PP_2. Hence, M is the mid–point of PP_2 having the coordinates

$$M\{(x_1 + x_2)/2, (y_1 + y_2)/2\}. \qquad (2.4)$$

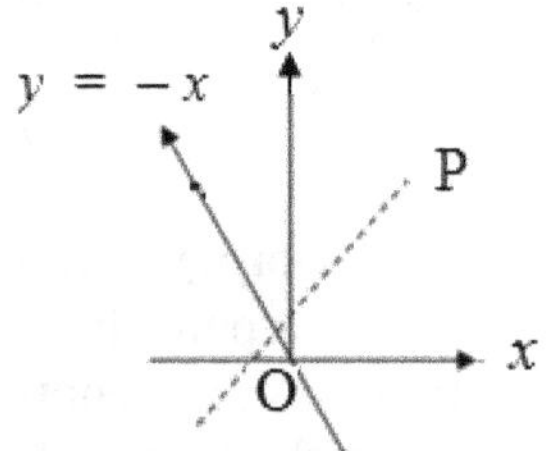

Fig. 2.8

Also, PP_2 being a line through P (x_1, y_1) with slope -1 has its equation

$$y - y_1 = -(x - x_1) = -x + x_1,$$

or,

$$y = -x + x_1 + y_1. \qquad (2.5)$$

Solving $y = x$ and (2.5) their point of intersection M also has the coordinates

$$M\{(x_1 + y_1)/2, (x_1 + y_1)/2\}. \qquad (2.6)$$

Comparing (2.4) and (2.6) we, thus, find $x_2 = y_1$ and $y_2 = x_1$. Thus, P_2 has the coordinates (y_1, x_1). So, in general, there holds the Eqn. (2.3). //

§ 3. Rotation

Rotation about a point, say O, through an angle θ is a transformation mapping a point P into the point P′ such that ∠ POP′ = θ and OP = OP′. It may be denoted by R_θ. Different values of θ can describe the same rotation. For example,

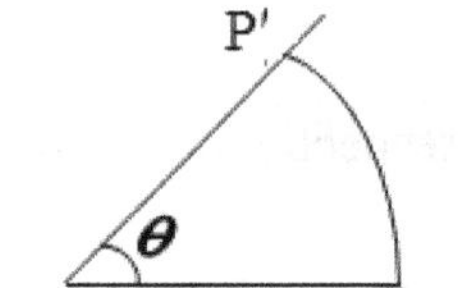

Fig. 3.1

$$R_\theta = R_{2\pi - \theta},\ R_{180°} = R_{-180°},\ \text{etc.} \qquad (3.1)$$

provided all these rotations have the same centre at O.

3.1. Centre of rotation

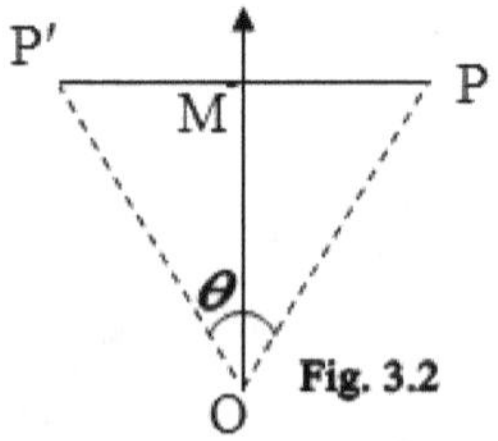

Fig. 3.2

If a point P maps onto P′ under a rotation then the centre of rotation O lies on the perpendicular bisector of PP′.

Definition 3.1. A figure is said to have *rotational symmetry* if it fits into itself at least once before completing a revolution.

3.2. Rotation in the coordinate plane

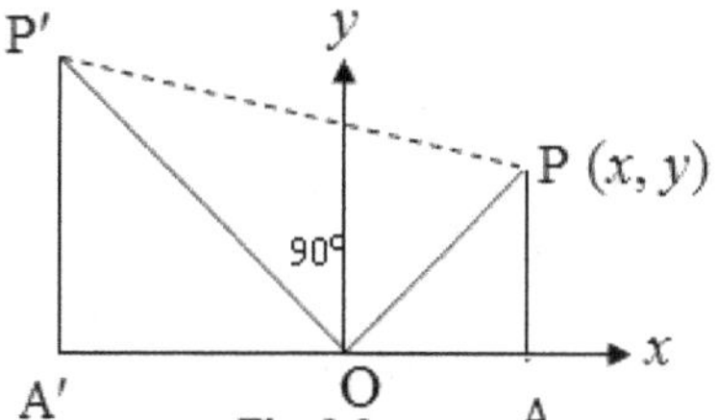

Fig. 3.3

Theorem 3.1. Under a rotation through 90° anti–clockwise about the origin O, a point P (x, y) maps onto the point P′ $(-y, x)$.

Proof. Drawing ordinates PA and P′A′ we construct two right triangles OAP and OA′ P′ which are congruent to each other. Therefore, OA′ = AP and A′ P′ = OA. Hence, the coordinates of P′ are (– OA′, A′ P′). //

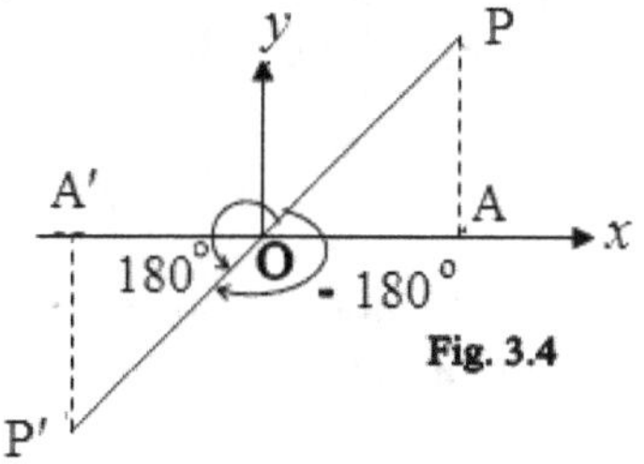

Fig. 3.4

Theorem 3.2. Under a rotation through 180° (clockwise or anti–clockwise) about the origin O, a point P (x, y) maps onto the point P′ $(-x, -y)$.

$$R_{180^\circ}(x, y) = (-x, -y). \qquad (3.2)$$

Proof. Drawing ordinates PA and P′A′ we construct two right triangles OAP and OA′ P′ which are congruent to each other. Therefore, OA′ = OA and A′ P′ = AP. Hence, the coordinates of P′ are (– OA′, – A′ P′).//

§ 4. Isometries

4.1. Transformations preserving congruence

Definition 4.1. Transformations preserving distances are called *isometries*.

Thus, if any two points P and Q are mapped onto P′ and Q′ respectively under an isometry then the distances PQ and P′ Q′ remain equal.

Example 4.1. Translation is an isometry.

Example 4.2. Reflection through a line is an isometry.

Example 4.3. Rotation about a fixed point is an isometry.

§ 5. Groups of Transformations

5.1. Product transformations

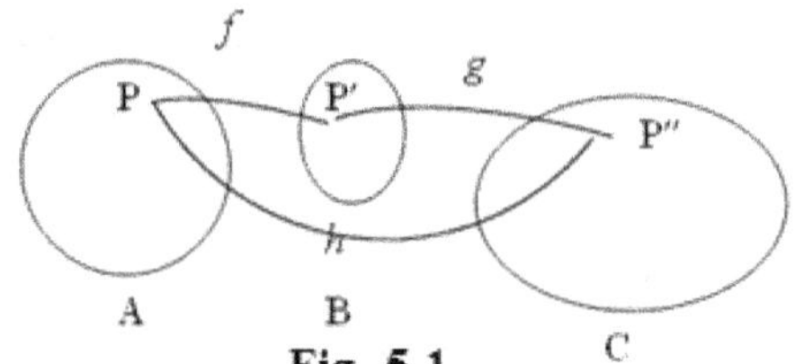

Fig. 5.1

Let A, B and C be any three geometric configurations and f transforms A onto B. Further, let another transformation g maps B onto C. The combination $h = g \circ f$ of maps f and g is defined as the product transformation if for each point P of A there exist P′ in B and P″ in C such that there hold

$$h\,(\mathrm{P}) = g \circ f\ (\mathrm{P}) = g\,\{f\ (\mathrm{P})\} = g\,(\mathrm{P}') = \mathrm{P}''. \qquad (1.1)$$

The product is defined in such a way that the transformation on the right, i.e. f, in the product is performed first. In Figure 5.1, we observe that

$$\mathrm{P} \xrightarrow{f} \mathrm{P}', \qquad \mathrm{P}' \xrightarrow{g} \mathrm{P}'' \qquad \text{and} \qquad \mathrm{P} \xrightarrow{g \circ f} \mathrm{P}''.$$

Thus, P, P″ form a pair in the product transformation h.

Example 5.1. Let $f = \{(3, 4), (2, 6), (7, 1)\}$ and $g = \{(4, 5), (6, 3), (1, 2)\}$ be transformations of points on a line, i.e. f maps the points 3, 2, 7 onto 4, 6,1 respectively and g further maps the points 4, 6, 1 onto 5, 3, 2 respectively. Their product transformation $g \circ f$ maps the points 3, 2, 7 as follows:

$$g \circ f\,(3) = g\ \{f\,(3)\} = g\,(4) = 5, \qquad g \circ f\,(2) = g\ \{f\,(2)\} = g\,(6) = 3,$$

and

$$g \circ f\,(7) = g\ \{f\,(7)\} = g\,(1) = 2 \ \Rightarrow\ g \circ f = \{\,(3, 5), (2, 3), (7, 2)\}.$$

We note that the product $f \circ g$ cannot be defined, for $f \circ g\ (4) = f\ \{g\ (4)\} = f(5)$; but, transformation of 5 is not defined under f. //

Example 5.2. Let f and g be two translations defined by

$$f(x, y) = (x + 1,\ y + 1) \text{ and } g\,(x, y) = (x + 2, y - 2).$$

Their product transformations $g \circ f$ and $f \circ' g$ map any point P (x, y) as follows:

$$g \circ f(x, y) = g\{f(x, y)\} = g(x + 1, y + 1)$$

$$= (x + 1 + 2, y + 1 - 2) = (x + 3, y - 1),$$

and

$$f \circ' g(x, y) = f\{g(x, y)\} = f(x + 2, y - 2)$$

$$= (x + 2 + 1, y - 2 + 1) = (x + 3, y - 1).$$

In this case, both product transformations $g \circ f$ and $f \circ' g$ are defined and are equal as either of these map the point P (x, y) onto the same point P′ $(x + 3, y - 1)$. The product transformations are also translations with tail at the origin and head at $(3, -1)$. //

CHAPTER 15

NON–RELATIVISTIC SCATTERING OF MATTER

§ 1. Introduction

The most important experimental technique in quantum physics is the scattering experiment. In atomic physics, Rutherford's discovery of the nucleons was based on his study of particle scattering off a gold foil, and the Frank-Hertz experiment established the existence of atomic energy levels by observation of electron scattering off mercury vapor. In nuclear physics, the first clear evidence of nuclear structure came from Rutherford's observation of scattering reaction:

$$\alpha + {}^{14}N \;\rightarrow\; p + {}^{17}O.$$

On the other hand, in the study of elementary particle physics, the scattering experiments not only provide the experimental data, but it is also the principal means for the creation of particles themselves, as for instance, in the pion - production process

$$p + p \quad\rightarrow\quad p + p + \pi^0.$$

The theoretical tool for the analysis of scattering experiments is the scattering theory whose discussion requires various classifications of the subject. There are relativistic and non-relativistic theories. Next, there are single channel and multi-channel scattering, which are briefly discussed here. In most collisions there are many different sets of particles emerging in the final stage. For example, when α – particles are fired at nitrogen some of the different possible configurations are

$$\alpha + {}^{14}N \rightarrow \alpha + {}^{14}N,\;\; \alpha + {}^{14}N \rightarrow p + {}^{17}O,\;\; \alpha + {}^{14}N \rightarrow \alpha + \alpha + {}^{10}B, \text{ etc.}$$

Each of the final sets of particles is called the *channel*. As such, this is a multi-channel process. However, certain single processes have only one channel. Two examples of such single channel processes are low energy scattering of electrons off protons or neutrons off α – particles. In either of these processes the only possible outcome in elastic scattering

$$e + p \rightarrow e + p \qquad \text{and} \qquad n + \alpha \rightarrow n + \alpha.$$

The concept of a single channel process is really an idealization. If, for example, we increase the energy of $n - \alpha$ collisions above about 20 MeV the neutron can knock the α -particle apart; and, in the $e - p$ example there is always the possibility of producing low-energy photons whatever be the incident energy. None of these examples are truly single-channel collision. Nonetheless, there are many processes that can be approximated as the single-channel collisions under the right conditions. So, within the framework of non-relativistic quantum mechanics the scattering of a single particle off fixed potential, and of two particles off one another provide completely consistent models of single-channel systems. Thus, the formulation of single-channel scattering is naturally much simpler than the multi-channel problem but at the same time it includes almost all the basic concepts needed for the later. Other main consideration of the scattering theory is its time-dependent or time-independent character. The first one deals with the time-dependent wave function describing the process of collision as it actually occurs. Well before the condition begins and after it is all over, the particles involved behave just like free particles; and, therefore, the corresponding wave functions behave like free wave functions. It results into a relation connecting the wave functions before and after the collision by a unitary operator S called the scattering operator. As a result, all experimentally relevant information is contained in operator S and the experimentally measured scattering cross section can be expressed in terms of matrix elements of S.

The time-independent formalism arises from an expansion of the actual time-dependent wave functions in terms of the so-called stationary scattering states, which are just the approximate *eigen* functions of Hamiltonian. The principal usefulness of this is to provide the means of actual computation of scattering operator (or related scattering amplitude) and for establishing its general properties. Thus, categorizing the scattering theory, we have

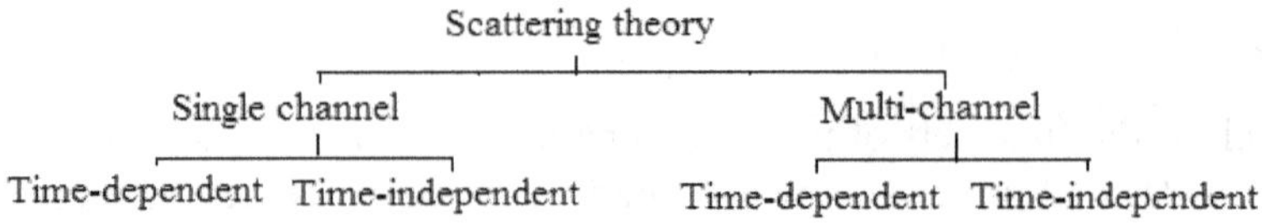

Historically, it is developed in analogy with the theory of bound states around the stationary scattering states; i.e. in terms of time-independent states formalism. Only latter, time-dependent theory was developed to provide proper justification. Therefore, in the scattering of a single particle off fixed potential a wave function $\psi^+_p(x)$ defined as solution of time-independent Schrödinger equation with the boundary conditions

$$\lim_{r\to\infty} \psi^+_p(x) = \{e^{ip\cdot x} + f(E_r, \theta)\, e^{2pr}/r\}/(2\pi)^{3/2}.$$

This wave function represents a steady (state) incident beam of particles with momentum p plus a spherically spreading scattered wave with amplitude $f(E, \theta)$ leading to the differential cross-section

$$\frac{d\sigma}{d\Omega} = \frac{\text{scattered flux / solid angle}}{\text{incident flux / area}} = |f(E,\theta)|^2 .$$

Although it leads to a correct result but is still less than satisfactory. A wave function depending upon the variable x should represent the state of one particle and not a beam of particles. Further, ψ^+_p being not normalizable, it cannot represent a state at all. Also, ψ^+_p being eigen function of the Hamiltonian corresponds to a steady state situation which is in contrast with time-dependent collision. It also ignores the interface of two axes.

All above objections can be ruled out and a justified conclusion can be derived if we build up a normalized time-dependent wave packet by superposition of wave function ψ^+_p with suitable momenta p. In a collision problem the energy is specified in advance and the behavior of the wave function at great distances is found in terms of it. This asymptotic behavior can be related to the amount of scattering of the particle by the free field. To achieve the idea, consider a simple 1-dimensional collision of a particle with square potential behavior $V(x)$ shown in the figure. The particle is approaching from the negative region of x-axis and is reflected by the barrier. The particle is always reflected (respectively transmitted) if its energy is lesser (resp. higher) than that of its top of the barrier. In quantum problem both the reflection and transmission occur with finite possibility for most energies of the particle. Due to lack of symmetry between positive and negative values of x we assume that

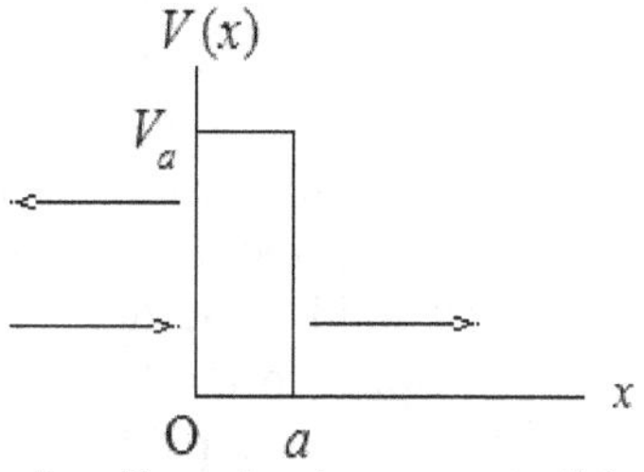

One-dimensional square potential behavior of height V_a and thickness a.

$$V(x) = \begin{cases} 0, \text{ when } & x<a, \text{ and } x>a \text{ ;} \\ V_a, \text{when } & x=a, \ V_a \text{ being positive.} \end{cases}$$

We are interested in representing a particle that approaches from left with energy $E > 0$, and turns back by potential barrier. Hence, its asymptotic behavior in the region $V(x) = 0$ is as follows:

(i) For $x < 0$, the wave function is desired to represent the *reflected* particle moving towards *left* while the *incident* particle moving towards *right*;

(ii) For $x > 0$, the wave function is desired to represent only a *transmitted* particle moving towards *right*.

A particle in a force – free region moving in a definite direction with a given energy has necessarily a definite momentum. Hence, it can be represented by a 1-dimensional momentum eigen function

$$U(x) \propto e^{ipk/h},\ x > 0; \quad \text{and} \quad e^{-ipk/h},\ x < 0,$$

where p is momentum. For increasing $E > V_0$ the transmission coefficient oscillates between a strictly increasing lower envelope and unity. There is a perfect transmission when $\alpha a = \pi, 2\pi, \cdots$; whenever the barrier contains an integral number of half wave-lengths.

Likewise, the collisions in 3-dimensions, where a particle collides with a fixed free field, or two particles collide against each other, can be worked out. The angular distribution of particles scattered by a fixed centre of force (or by other particles) is conveniently described in terms of a scattering cross-section. Let us bombard a group of n particles or scattering centres with a parallel flux of N particles per unit area and per unit time and count the number of incident particles emerging per unit time in a small solid angle $\Delta\omega_p$ about a direction having polar angles θ_0 and φ_0 with respect to the bombarding direction as the polar axis, the number will be proportional to N, n and ΔN_0. So, the number of incident particles emerging per unit time in $\Delta\omega_p$ is $nN\sigma_1(\theta_p, \varphi_p)\ \Delta\omega_p$, where the proportionality factor $\sigma_1\ (\theta_p, \varphi_p)$ is differential scattering cross-section. Flux being small there is no interference amongst bombarding particles and no appreciable diminution of bombarded particles by their recoil out of target region. Since $nN.\sigma_0\ (\theta_0, \varphi_0).\ \Delta\omega_p$ is of dimension reciprocal to time, $\sigma_0\ (\theta_0, \varphi_0)$ has dimension of an area. Therefore, total scattering cross-section

$$\sigma_p\ (\theta_0, \varphi_0) = \int \sigma_0\ (\theta_0, \varphi_0)\ d\omega_p$$

over the surface of sphere. Since the two particles move in opposite directions away from each other after collision, it is clear that the differential cross-section for observation of the recoil of bombarded particle in the direction (θ, φ) is just $\sigma\ (\pi - \theta, \varphi_\alpha + \pi)$.

BIBLIOGRAPHY

1. Grewal, B. S. (1998): *Higher Engineering Mathematics*, 34th edition, Khanna Publishers, India.

2. Kreyszig, E. (1993): *Advanced Engineering Mathematics*, 7th edition, John Wiley and Sons, USA.

3. Misra, R. B. (2002): *Sadish evam unke Anuprayog* (Hindi translation of "A course in Vectors and their Applications" by Prof. R.S. Mishra), Hardwari Publications, India, pp. x + 244, ISBN 81-88574-02-3.

4. Misra, R. B. (2010): *A Text-book of Classical Mechanics*, Lambert Academic Publishers, Germany, ISBN 978-3-8433-8306-6.

5. Misra, R. B. (2010): *Basic Mathematics at a Glance*, Lambert Academic Publishers, Germany, ISBN 978-3-8433-8696-8.

6. Misra, R. B. (2010): *Complex Analysis*, Lambert Academic Publishers, Germany, ISBN 978-3-8433-8859-7.

7. Misra, R. B. (2010): *Laplace Transform, Differential Equations and Fourier Series*, Lambert Academic Publishers, Germany, ISBN 978-3-8433-8328-8.

8. Misra, R. B. (2010): *Theory of Sets, Groups, Rings, Fields, Integral Domains, Vector Spaces, Metric Spaces and Topological Spaces*, Lambert Academic Publishers, Germany, ISBN 978-3-8383-9943-0.

9. Misra, R. B. (2010): *Transformation Geometry*, Lambert Academic Publishers, Germany, ISBN 978-3-8433-8827-6.

10. Misra, R. B. (2011): *Advanced Integral Calculus*, Lambert Academic Publishers, Germany, ISBN 978-3-8443-1916-3.

11. Misra, R. B. (2011): *Engineering Mathematics*, Lambert Academic Publishers, Germany, ISBN 978-3-8433-8931-0.

12. Misra, R. B. (2011): *Glossary of Mathematics*, Lambert Academic Publishers, Germany, ISBN 978-3-8443-0203-5.

13. Sinha, R.S., and Srivastava, B. K. (2005): *A text-book of Differential Equations*, 2nd edition, Chandra Prakashan, India.

14. Spiegel, M. R. (1965): *Schaum's Outline of Laplace Transforms*, Mcgraw-Hill, USA.

15. Stroud, K.A. (1990): *Further Engineering Mathematics*, 2nd edition, Macmillan Publishers, Ltd., UK.

INDEX

E

F

G

H

I

K

L

M

N

O

www.ingramcontent.com/pod-product-compliance
Lightning Source LLC
Chambersburg PA
CBHW071339210326
41597CB00015B/1502
* 9 7 8 1 9 2 5 8 2 3 1 1 0 *